COMMUNITY RESILIENCE IN
NATURAL DISASTERS

COMMUNITY RESILIENCE IN NATURAL DISASTERS

Edited by

Anouk Ride
and
Diane Bretherton

palgrave
macmillan

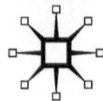

COMMUNITY RESILIENCE IN NATURAL DISASTERS
Copyright © Anouk Ride and Diane Bretherton, 2011.
Softcover reprint of the hardcover 1st edition 2011 978-0-230-11428-9

All rights reserved.

First published in 2011 by
PALGRAVE MACMILLAN®
in the United States—a division of St. Martin's Press LLC,
175 Fifth Avenue, New York, NY 10010.

Where this book is distributed in the UK, Europe and the rest of the world,
this is by Palgrave Macmillan, a division of Macmillan Publishers Limited,
registered in England, company number 785998, of Houndmills,
Basingstoke, Hampshire RG21 6XS.

Palgrave Macmillan is the global academic imprint of the above companies
and has companies and representatives throughout the world.

Palgrave® and Macmillan® are registered trademarks in the United States,
the United Kingdom, Europe and other countries.

ISBN 978-1-349-29585-2   ISBN 978-0-230-33932-3 (eBook)
DOI 10.1057/9780230339323

Library of Congress Cataloging-in-Publication Data

Ride, Anouk.
    Community resilience in natural disasters / Anouk Ride,
Diane Bretherton.
        p. cm.

        1. Disaster victims. 2. Disaster victims—Psychology. 3. Disaster relief.
    4. Helping behavior. 5. Solidarity. 6. Communities. I. Bretherton, Di. II. Title.

HV553.R525 2011
363.34'81—dc22   2011007026

A catalogue record of the book is available from the British Library.

Design by Newgen Imaging Systems (P) Ltd., Chennai, India.

First edition: August 2011

10 9 8 7 6 5 4 3 2 1

Transferred to Digital Printing in 2012

CONTENTS

ACKNOWLEDGMENTS

This book began with conversations in Mexico City where people recalled their stories of resilience in the aftermath of one of the country's worst disasters. Recalling the experience of 1985, people remembered the losses and the suffering but at the same time the word that kept coming up was: solidarity.

As Mexican Disaster Planner and Educator Graciela Zapata Sierra outlined, this was a source of pride:

> In other countries, I have not seen the kind of cooperation we have here when a disaster happens in Mexico. For example, we think that 85 percent of the accidents are attended by volunteers. Why do we have this here? I think its culture. When the earthquake happened I was only 12 years old so I could only help my family. Had I been in secondary school I would have helped others. I don't know why Mexican people want to help but I think its culture.

According to Enrique Ortiz, director Habitat International Coalition Mexico, the 1985 earthquake had a transformative effect as homeless people demonstrated their collective strength to the authorities:

> Literally a week after the earthquake residents were protesting and what is special is that they won. The sensitivity of the Mexican people to earthquake victims and the involvement of the World Bank in funding reconstruction meant the government had to act. This process did not start with the earthquake but the earthquake accelerated the process of people organising for housing rights. More recently, developers have taken over again and much of the problems remain but the networks and organizations that grew in those times remain.

As Fernando Álvarez Bravo, Brigada de Rescate Topos Tlaltelolco, (the rescue brigade of moles of Tlaltelolco) said,

> The earthquake changed the way the government saw the people of Mexico City. The government saw that if they don't do things the

people know they can do it alone, they can organise themselves and they don't need to wait for the government before acting.

Reflecting the popular feeling of camaraderie in Mexico City, the Solidarity Gardens were built over the site of Hotel Regis that collapsed in the center of the city during the earthquake—an example of a collective memorial, usually reserved for soldiers and revolutionaries, for the average citizen.

The people of Mexico City started a conversation between us, the authors, about things we had seen as we worked as journalists, educators, and peace and conflict researchers in different parts of the world. We had seen that the media and aid agencies were telling us a familiar story of disaster victims and social disorder and that community members were not able to tell us their stories through the usual channels of power. We saw that in the worst cases the authorities were even resisting signs of community resilience, such as Rodolfo Mora Rodríguez, from Section 15 of Hospital Juárez described about a disaster site where local people were rescuing and giving medical treatment to people trapped under the earthquake rubble:

> The army came and all of us civilians were kicked out; we were told that they had trained people. We couldn't go on with the job...they cordoned off the entire disaster area and they didn't let us do a blessed thing but they didn't do a blessed thing either....In a flash we got a little rally together. We started to say the authorities were to blame because our *compañeros* [comrades/friends] were still underground and we weren't allowed to go to their aid...We told [the commander] that instead of machine guns all his people should be carrying shovels no? So in the end they allowed us to go in. (Poniatowska, 1988: 23)

We wanted to demonstrate community resilience where it exists and explore how this can be recognized by outsiders, government, and non-government agencies. This endeavor would not have been possible without the willingness and time people gave to speak about their experiences facing natural disasters. We are also thankful to our researchers for establishing the trust and understanding needed to uncover these community stories of survival and resilience. Many people involved in this book, as researchers and interviewees, have personally been affected by disasters and then have gone on to help others. We want to acknowledge the contributions to this book from all those local people who figuratively and literally went into the disaster sites with shovels.

REFERENCE

Poniatowska, E. (1988), *Nothing, Nobody: The Voices of the Mexico City Earthquake*, Philadelphia: Temple University Press.

1

INTRODUCTION

As Mexico City woke up at 7.19 A.M., on September 19, 1985, an earthquake of 8.1 on the Richter scale shook the metropolis for three minutes. At first everything was quiet, then people began to scream and call out to one another—the shocked, the injured, and the mournful.

Already believing Mexico City to be a dangerous, poor, insecure, and chaotic place, the Western media spread to the world these moments—the discovery of bodies crushed by debris, the faces of the traumatized, and the damaged buildings that dotted like crumpled pieces of paper through the cityscape. Viewers felt pity and compassion for the people and a sense of distress and fear that a sudden force of nature could stop one of the world's largest cities.

In those three minutes of the earthquake, about 5,000 were injured with 1,000 trapped under the debris. Approximately 400 buildings were destroyed with over 3,000 damaged, as were 95 of the city's murals. Much of the city had no power, all of it had no telephone service, and in 60 percent of the city there was no drinking water. An estimated over 200,000 people were made homeless (Cessaro and Romero, 1987).

After those moments of disbelief, awe, and horror, something else happened. Something that remained largely hidden from the media, from the aid agencies that rushed to book flights into Mexico City, and from the national government. Something happened that would have escaped attention entirely had it not been so unexpected and so undervalued. People began to act.

Survivors called out and identified their family members, workmates, schoolmates, and neighbors who were trapped under the rubble, quickly forming groups of people to lift, piece by piece, chunks of cement and bricks to try and reach them. Volunteer rescuers, who came to be known as "moles," formed teams of people who followed the direction of locals to go into the smallest spaces to try and rescue survivors.

Roberto C. Hernández Alarcón, president of one of the "mole groups," Brigada de Rescate Topos Mexico, elaborated thus on the events in Ttatleloco, where 192 apartments in one building collapsed:

> In Mexico we talk to people on the street, we get to know our neighbours, people know everything about you, where you are going, what you are doing, when are you coming back. This is the way we make friends. So, when the buildings fell down in Ttatleloco, everybody knew where people were trapped and wanted to help get them out.

When word went out that rescuers lacked equipment, picks, shovels, and torches began to arrive—given by stores, dug out from people's garages, and delivered by crews of road workers, mechanics, boot makers, and many others.

Roberto Chavez Manjarrez, chief director, Prehospitalisation and Operations, Cruz Roja Mexicana (Red Cross Mexico) reported:

> By the time the Red Cross got to the disaster sites, people were already there—some had gone in to rescue people already, others were asking "how can we help"? As we worked at rescuing people, people brought bread and water for the victims and for us because they knew we had been working hard. Some people even came down just to encourage the rescue workers, shouting out things like "thank you!," "you can do it!"

At the Hospital General, the city's cheapest hospital, and the Hospital Juarez, doctors and nurses, forced to evacuate their crumbling wards, set up makeshift tents outside and began treating injured people. Then, in the next few days, new initiatives emerged—the university students and the Salvation Army set up camps where the rescuers could sleep for a few hours before beginning rescue work again, which continued night and day for weeks (Poniatowska, 1988: 211). Artists such as Manuel Lara visited camps of homeless people, playing painting games to entertain the children. Soap opera actresses performed people's favorite episodes and encouraged the women to work together to keep the camps clean and safe. The Psychoanalytic Association of Mexico organized initial group-therapy sessions in homeless shelters, which included identifying a key homeless person in each group to organize regular meetings, so that groups could become self-organizing support networks.

Within moments, building into hours, days, and weeks of cooperation, the city that had been shaken by the earthquake for a few minutes was back in the hands of the people, not just in a practical sense but in a psychological one as well, as pointed out by the president of the Mexican Association of Child and Adolescent Psychiatrists, Dr. Eduardo de la Vega:

> There is a lot of spontaneity, helping with rescues, directing traffic, finding people places to stay. It is helpful in itself and it is very useful in working through some of the feelings of impotence and guilt in this situation. (as quoted in Treaster, 1985: 8)

Why was this level of cooperation surprising? It was a demonstration of communities—normally invisible to many powers that be—becoming visible by their actions in a time of crisis. The fact that Mexico City had high rates of poverty and crime had made these communities perhaps even less visible than usual.

Talking to people in Mexico City about this disaster prompted our thinking about why people have such a negative view of how people react to disasters. In turning to the scholarly literature on natural disasters, we found it tends to take the viewpoint of the outsider. It sees the expert as objective and distant from the lives of the people affected by natural disasters with knowledge that is applicable to many, if not all, places. Researchers in this field tend to assume that the community is a preexisting entity, one that needs to be educated and otherwise changed to mitigate future hazards, risks, and vulnerabilities to natural disasters and their effects. The people who suffer the most from disasters are often marginalized people—not the academics who write papers or media commentators who bring reports to the public. Some cultures are much more oral than others and have less voice in publications. Marginalized groups may also come from language groups that are not dominant and may be excluded even from media that do not require writing, such as radio. They may also be politically repressed and silenced, for their voice might be critical of the prevailing arrangements. However, this does not mean that marginalized communities are without agency.

So, where could we look for an alternate view of communities facing natural disasters? One that recognized that people could determine their own future through their thinking and actions?

To begin to answer these questions, we sought studies to clarify our conceptual framework—specifically about what makes up communities and how they function. As a starting point, we look at the world of people as similar to and interlinked with ecology, with a focus on community as the level of analysis. Human relationships are envisaged as a set of systems, nested within larger systems, which are themselves nested within yet larger systems. Bronfenbrenner (1979) likened this idea of systems within systems to a set of Russian dolls, but of course the reality is that the systems do not stack as neatly inside each other as do the dolls, but rather constantly evolve and change their shape. We can also see that each environment is different and that microsystems can vary across cultures. There are the expected ones, like family, class at school and local church, monastery or mosque, but there are also other systems such as the tribe or "one language" groups, social networks such as sporting and literary associations, and many others used to organize community responses.

The network of human relationships exists within the physical environment, which is also complex and systemic. Following the usage of the Resilience Alliance, a group of researchers interested in resilience, we call this complex social and physical system "socio-ecology." To study the systems in the socio-ecology, it is possible for the observer to zoom in on local interactions or conversely to move back and take a wider view of how local events impact upon, or are impacted upon by, the wider social and physical systems. The point from which events are observed can also be varied: Is the observer outside the system, looking in, or is the observer a participant, an insider, experiencing and observing events as they unfold, trying to make sense of the flow of experience and deciding how to respond? A full understanding will need to consider both viewpoints, the "close up" of engagement and participation of a community involved in a disaster, and the outsider view that can allow for learning through contrast and comparison.

So, focusing our analysis on community, what do we mean by community? A community is a social network, rather than an organization that has a central authority and clear boundaries of inclusion and exclusion. Some scholars and policy makers classify these networks and bonds between people as social capital. Robert Putnam popularized the term "social capital" as: "features of social life—networks, norms and trust—that enable participants to act together more effectively to pursue shared objectives" (Putnam, 1996: 56).

Implicit in these frameworks, and their application by organizations such as the Organisation for Economic Cooperation and Development (OECD) and national governments, is the concept of the monetary value of social capital and "strong communities" in terms of needing fewer crisis services, creating economic value through relationships, reduction or lack of crime/violence, and reduction of the number of people who are socially isolated and dependent on the state or aid agencies for support. Coleman (1988) and Putnam (1993, 1996) recognized social capital could be good or bad in consequence. Recent research has further classified links in communities into two types of social capital: "bonding" (links between people of similar class, ethnicity, etc.) and "bridging" (links between people of different class, ethnicity, etc.). Government interventions in communities aimed at social inclusion will often seek to create or increase bridging social capital (OECD, 2001).

So, the nature and culture of a community is like an organism that grows with the communication and interaction of its members. Over time it might reach out to incorporate new members, or cold shoulder and edge out old ones. As there is no central authority or a written code of conduct, vertical rather than hierarchical communications are important. That is, people communicate with and learn from each other. The learnings from these interactions—shared values and beliefs and expected ways of behaving—are said to be the culture of a community. However, a community is also diverse, with different groups of people taking up different roles and responsibilities. A community encompasses the idea of being part of a group, of belonging. Membership is not necessarily formally conferred, but may be achieved by shared communication and working together. In this sense, a vibrant community might be said to have a "spirit."

If we look at a group of people and suppose them to be a community, we might look at a number of dimensions. Is there a connection between people? Are there shared norms that influence behavior? Do people communicate and cooperate with each other? Do people interact creatively and adapt to changing circumstances? Do people have the capacity to work with and transform conflicts?

Second, we come to the idea of resilience—how can we say if a community is resilient? What are the defining characteristics of resilience? What does it look like?

A core theoretical idea is that, when threatened by negative change, systems may vary in their degree of resilience. The *New Shorter Oxford Dictionary* defines resilience as:

1. The action or an act of rebounding or springing back.
2. The ability to recover readily from or resist being affected by an illness, setback, etc.

The Resilience Alliance researchers define resilience as the capacity of a system to tolerate disturbance without collapsing into a qualitatively different state that is controlled by a different set of processes. A resilient ecosystem can withstand shocks and rebuild itself when necessary. Resilience in social systems has the added capacity of humans to anticipate and plan for the future. Resilience as applied to ecosystems, or to integrated systems of people and the natural environment, has three defining characteristics: (1) the amount of change the system can undergo and still retain the same controls on function and structure, (2) the degree to which the system is capable of self-organization, and (3) the ability to build and increase the capacity for learning and adaptation.

While the resilience literature acknowledges the equal importance of social and physical realms, its research papers more frequently deal with environmental conditions and indicators. Our research differs in focusing more particularly on the characteristics of social groups—communities—and their capacity to withstand shock or natural disaster. We will draw on the work of the resilience researchers insofar as complex systems theory applies to social as well as ecological systems.

Resilience is a term used to describe qualities of individuals, communities, and societies, but community resilience refers to not just a collection of resilient individuals but a collective state, as Brown and Kulig (1996/97: 93) said: "People in communities are resilient *together*, not merely in similar ways." However, defining community resilience has proved difficult. Community resilience has variously been described as an ability (Brown and Kulig, 1996; Paton and Johnston, 2001; Pfefferbaum et al., 2005), process (Sonn and Fisher, 1998), development of resources (Ahmed et al., 2004), sense of ability of the community (Shamai and Kimhi, 2004), and a set of capacities, skills, and knowledge (Coles and Buckle, 2004). Resilience definitions have emerged from the idea that communities might return to a static state to the idea that resilience is not "bouncing back" but "bouncing forward" to a new state where weaknesses underlined by the disaster are acknowledged and acted on. Definitions have also placed an increasing focus on not just coping and recovery but also on adaptation. Synthesizing various definitions, Norris et al. defined community resilience as "a process linking a network of adaptive capacities to adaptation after a disturbance or adversity" (Norris, Steven, Pfeffererbaum et al., 2008: 137). Communities may be described as resilient then if they have shown or are showing "a positive trajectory of functioning and adaptation after an initial disturbance as a result of adequate adaptive capacities" (Norris, Steven, Pfeffererbaum et al., 2008: 131). Community resilience is transformational rather than referring to strengths such as social capital and community competence.

Resilience, as we see it applied to natural disasters, is the capacity of a community to cope with the emergency, to rebuild, and to learn from the experience, such that the new physical, social, and political structures are better adapted to the environment. So, we can look at why some communities seem to be resilient and others are more fractured and disturbed by the disaster that they might seem dysfunctional or unable to adapt to the postdisaster reality.

Third, we looked to see why our interest, as peace and conflict researchers, had been peaked by natural disasters. After all, what has peace got to do with natural disasters?

At first, the idea of peace and that of disaster relief may seem like distant concepts, often the former being seen as a state of being between adversaries or combatants and the latter being seen as a provision of essential goods from the fortunate to the needy. However, both conflict and natural disasters are crises for communities that can both reveal and shape a community's resilience and prospects for peace.

Disasters are interesting` because they tell us something about human behavior. The disaster calls upon communities to respond, and exploring these community responses provides information about how a community functions, treats outsiders, and rises to new challenges. Community responses also inform us about cultures, the habitual ways of doing things, and the meaning ascribed to these patterns of action, about what is "normal" and "abnormal" behavior in a given community, and of how this can be stretched to its outer limits by the circumstances. By its very nature, a disaster brings out the extremes of human experience, and explanations for these extremes; of how communities and nature are connected, about faith, and fate.

Our research stems from an interest in resilient communities and a commitment to the values of peace and the nonviolent resolution of conflict. Early peace researchers, like Galtung (1978), pointed out that peace for many years lacked definition and was seen as a state of "not war" more than possessing positive attributes. Following the opening up of this line of inquiry, feminist writers like Elise Boulding (1990, 2000) elaborated on the concept of peace by noting that peace is not just the business of military leaders and heads of state but is also a feature of more everyday settings, such as the home, school, workplace, and community. In this way, peace involves all people, not just those formally engaged in decisions to use armed force. Instead, community members can contribute to a "culture of peace" enabling harmonious relationships and respect between different peoples.

Aid and nongovernment agencies are increasingly concerned that projects contribute to, or at least refrain from undermining, peace-building efforts (the "do no harm" approach popularized by the book by the same title in 1999). It is wearying for agencies to put resources and efforts into development and then find their efforts are undermined by civil strife. It is even more galling to recognize that sometimes efforts to assist other countries can undermine local structures and increase, rather than decrease, security. There is also the fact that the majority of natural-disaster deaths and disaster-related illnesses occur in the developing world, and many disasters also occur in conflict-affected countries experiencing higher rates of poverty.

However, whereas there is increasing concern about the need for peacebuilding in aid and development, there is not a clear articulation of how peacebuilding can be measured and assessed. Monitoring and evaluation by aid agencies tend to only go so far: whether the project has met its set goals and not whether the concerted efforts of aid agencies are having a positive or negative impact on prospects for peace in the areas where they work.

But how do we know when peace has been built? Societies may appear peaceful on the surface but be quite fragile. For example, if neighbors turn on one another rather than cooperate when stress occurs then this might indicate that the prior peaceful state was only negative—"not war" but also not peace in the way modern thinkers have come to understand a state of positive peace.

A key feature of social capital is that it is built through networks based on regular interaction between individuals and families. In this way, services that interact with regular life such as schools, kindergartens, health clinics, hospitals, and welfare agencies have become seen as an important means to build social capital such as through trust and social-support networks between disparate people. In the context of a natural disaster, relief services become connectors bringing together different survivors and service workers that can create bridging social capital or alternatively can override and weaken social connections and so expose the community to conflict in the future.

In this and many other ways, culture (including a culture of peace) can be impacted by natural disasters. Literature on disasters tends to be written from a universalist perspective, and the relationship between culture and disaster management is sparse. However, the duress of the disaster situation will both evoke and disrupt culture, including gender roles and relations. On the one hand, shared habitual ways of behaving will enable communities to communicate and work together quickly and effectively. On the other hand, the disaster has the potential to disrupt these habitual ways of behaving. As fundamental aspects of culture are tacit and taken for granted, they dwell below the level of people's awareness. Natural disasters may bring these aspects to mind, enabling cultural change and adaptation.

Peace researcher and conflict-resolution practitioner John Paul Lederach (1997, 2005) suggests that the root of culture is in meaning. The need to interpret and make sense of events will be evident in a disaster situation, and the different cultural groups will bring differing frames of reference to this task. A natural disaster will function to draw people from different cultures together and create cross-cultural exchanges. Immediately after the disaster, local communities will cope as best they can to survive. When outside help begins to arrive from the national authorities, perhaps in the form of the army or officials from the capital city, different cultural expectations will be introduced. As international humanitarian assistance begins to arrive, it too begins new cultural assumptions and ways of operating. These various cultures may find it difficult to cooperate effectively with each other in responding to the natural disaster.

Just as disaster management literature can be blind to culture, it can be a "top-down" "experts-to-lay-community members" approach and undermine the agency of local communities. Structuration theory, as described by Vivienne Jabri (1996), explores the links between structure and agency, noting that while social actors are constrained by social structures they also have a role in reproducing or changing them. A key feature of structuration theory is that it is not deterministic and allows that social actions may have unintended effects. For instance, the Interagency Standing Committee of the UN (1999: 1) notes that aid can not only be more efficient but will also have greater impact if the "opportunities for positive change in gender roles created by crisis situations are enhanced and sustained."

Seeing communities as capable of leadership in times of a crisis is not always easy for decision makers in governments and international organizations. Even for those who see the value in this approach, it is often difficult to do this on a practical level—especially when the perceived need to act quickly is heightened by the number of people in distress and need.

However, the benefits of peacebuilding outweigh its complexities. If people are able to cooperate and act, lives and health can be saved in the immediate response to a natural disaster. With the

United NationsDevelopment Programme's (UNDP) Bureau of Crisis Prevention and Recovery reporting that an estimated 75 percent of the world's population live in areas affected by an earthquake, flood, tropical cyclone, or drought from 1980 to 2000, the economic and political benefits of supporting community resilience are only becoming more obvious with each new natural disaster event. For the communities affected, their leadership and action during natural disasters also have a vital psychological role of reaffirming their belief in their own capacity as well as leading to new learning and possibilities.

Internationally, if the scientific forecast of an increase in extreme weather events is correct then perhaps fatigue will set in, compassion will wear thin, and even extreme events will not be dramatic enough to evoke a generous response from the international community. If aid agencies are reliant on the emotions of donors they may find it difficult to fund sustained efforts to build more self-reliant communities, efforts that are less dramatic than crisis intervention, but are in the long term more effective. So, peacebuilding after natural disasters has many imperatives, from political to economic, global to local.

Peace and natural disasters are linked at each stage of a community's response to the crisis. The response of a community to a natural disaster can become a "peace litmus test" to indicate the level of positive peace in a community, for it tells us how a community deals with crisis—whether peaceful cooperation between community members is evident, whether the community is inclusive or exclusive with regard to others in need, whether there are underlying disagreements that newly surface at a time of stress, and whether the community can adapt and learn from the experience.

Large disaster relief and development programs tend to come from international organizations and so from outside a local community. The impact of international assistance can add acid or alkaline to these positive peace levels in a community—the way assistance is communicated, formulated, distributed, managed, and assessed can impact a community's sense of agency, internal relations between different community members, relations between the community and the government, and perceptions of "outsiders" and the international community.

To explore all these issues, we elaborated a research methodology that would examine what is normally unexamined—how people inside the disaster made sense of what was happening and formulated a collective response (see Box 1.1: METHODOLOGY). Key to the research method was interviews, speaking with local people, and then

Box 1.1 Methodology

An approach used in the disciplines relevant to the study of community resilience in natural disasters—including psychology, development studies, peace studies, and social sciences—is the enhanced case study, an interpretative and analytical use of case studies through a conceptual framework. This allows for collection of data and analysis in different countries that can be compared and contrasted to support broader arguments about the behavior and action of people in response to events.

As our intent is to build an evidence base for community resilience in natural disasters in different cultural, environmental, social, and economic settings, the enhanced case study approach using most different systems design was selected to provide the first global study of the subject matter.

Two specific data-collection methods were used for these case studies: (1) a brief literature review (of both scholarly works and infield reports in aftermath of the natural disasters) to set the context for each case study and (2) infield interviews with informants to provide the basis for analysis of community action and behavior during the natural disaster and aftermath.

This dual approach of synthesizing preexisting literature and undertaking new research by interviewing people from the community provides a comprehensive study of the community while also allowing for tension between how the disaster is remembered in written record and the oral narratives of the community members themselves. For instance, the mainstream global media often depicts "victims" as helpless and reliant on outside actors to initiate recovery whereas community narratives are often different and can reveal much about cultural and social mores that contribute to community resilience in the absence of external support.

It is our intent that using this methodology will:

- Bring the voices of communities to the fore of this analysis of natural-disaster response and recovery, including women and disadvantaged groups.
- Identify, through the comments and reflections of community members directly involved in natural disaster response and recovery, strategies and actions that build community resilience.
- Uncover, through this direct experience of community members, strategies and actions that national and international organizations take to support or detract from community resilience.

Selection of Case Study Locations

In the selection of case study locations, we sought to represent:

- Various types of natural disasters (floods, hurricanes, earthquakes, etc.).
- Various levels of economic, social, and environmental destruction caused by natural disasters.
- Different levels of economic development.
- Various cultural, religious, and social settings.

With these factors in consideration, the final selection of case studies was made with a view to the availability of researchers with research skills, thorough knowledge of the case study location (including links to local NGOs and ability to speak local languages), and the ability to write well in English.

Selection of Sources

Informants who represent civil society in the case study location were interviewed by partner organizations to collect stories and direct quotes from local communities regarding their response to the natural disaster.

Informants were selected based on their connection with civil society networks in the community. Care was taken to attend to gender relations in the choice of researcher, informants, and questions asked.

An average of 11 informants from each location were selected to portray the community. Informants included the following:

- Religious representatives.
- Local government representatives.
- Civil society organization representatives (including women's networks, cultural and arts networks, sporting and social club networks).
- Relief and rescue workers (including volunteer and paid rescuers, health workers, and social workers).
- Educators, youth workers, and youth representatives.

Where informants contested information in the literature, this conflict is highlighted in the analysis, and where the majority of informants agree this information is incorrect, this was adjusted. In this way, the literature review and the interview research are dynamic in the research process and integrated in the findings.

For the discussion of community response and role of assistance, research was conducted in local languages with the assistance of researcher/s in each case study location.

A set of open-ended questions was provided to the researchers with the understanding that discussion may roam outside these questions, and this discussion was also valuable as an insight into the full explanation of events. A full transcript of each interview was provided to form a data set for analysis in consultation with the researchers.

Interviews

The interviews were conducted in the various countries by researchers with some knowledge of the local context, culture, and language. The interviews took a narrative approach and encouraged informants to tell the story of their experience of the disaster, allowing events to unfold over a time sequence, and then finally asking them to reflect on that experience. This involved setting up a supportive and empathic relationship, in the context of the relationship, drawing out ideas that had not previously been articulated. The interviews are not only quite therapeutic in the retelling of stories, but are also quite challenging in posing questions about the meaning of this experience. The interviews became a dialogue rather than an interrogation and are marked by their warm and constructive tone. Having played an important role in the response to the disaster, people were pleased to be asked for their opinions and ideas.

The final chapter of the research findings builds on comparative findings to draw on cross-country similarities and differences in the case studies to present findings on how communities act in times of crisis. These enhanced case studies are integrated with the broader, thematic literature reviews on responding to disasters to posit general principles to guide interventions toward strengthening local community resilience in natural disasters.

studying these with the assistance of a computer program designed to analyze text and transcripts.

We are interested in drawing on local knowledge, on the experience of communities that survive disasters and work together to recover and rebuild, because this voice is less well represented in the literature on disasters. Our methodology sees a community not as a fixed entity but as being created through interaction and able to learn through experience. Our research sees the local communities and the individuals who compose them as actors who have agency to adapt to new conditions and shape their own social arrangements.

The aim of our research is to better understand the experience of local communities so as to try and find ways in which outsiders can respect and build on local strengths to assist communities that experience natural disasters.

Our book is one of voices of people in the eye of the storm, the height of the water, and the dust of the drought—people who have experienced disasters, led community efforts, and emerged on the other side to reflect on how people survived, coped, and adapted to the crisis. We talk to them about what they think made communities resilient amidst extraordinary odds. This resilience is about more than "bouncing back"; it is about using the crisis for learning, change, improvement, and understanding.

In this way, community resilience can be envisaged not just as a piece of elastic that when stretched bounces back to its previous state, but more like a piece of fabric that grows, softens, and becomes larger and more flexible with the stretch. The disaster may pull, twist, and fray communities, putting stress on the fabric of society, which can seem to be on the brink of tearing apart. So how do communities respond to the pressure of the disaster? And how does aid and assistance from outsiders influence the community response? As the stories from people inside disasters demonstrate, even more dramatic than what a disaster can do to a community is what the community, faced with disaster, can do to itself.

Box 1.2   What Is a Natural Disaster?—A Definition

The critique of the term "natural disaster" reflects more holistic and systemic approaches to viewing environmental system disruptions such as disaster events. A natural disaster inevitably involves an interaction between human interaction, human vulnerability, and nature, rather than being entirely "natural." For example, Mexico City was originally built on an island of Lake Texcoco, now submerged, creating soft soils. The area has long been known to be subject to seismic activity, yet urban development proceeded with few plans and poor safety standards. Consider the view of Mexican volunteer Judith Garcia:

> People who died in 1985 didn't die because of the earthquake; that is a lie. People died because of poor construction, because of fraud, because of the criminal incapacity and the inefficiency of a corrupt government that doesn't give a damn about people living and working in buildings that can collapse. (quoted in Poniatowska, 1988: 93)

We define a "natural disaster" as the consequence of a natural hazard such as a volcanic eruption, earthquake, fire, flood, or landslide that affects human activities. An event such as an earthquake will not be deemed a disaster if it occurs in a remote area and does not affect human lives.

This study also explores whether "natural disasters" are seen as natural in the eyes of communities experiencing them. Evidence suggests that peoples' thoughts and feelings about events that are "acts of God" such as earthquakes, where people to blame are not easily identifiable, are distinct from thoughts and feelings about events that are "acts of humans" such as a war or a chemical spill, where people to blame are easily identifiable. Indeed, elements of community resilience may seem to be stronger in response to natural disasters than to other types of crisis such as a violent conflict. The perception that the disaster cause comes from outside the community helps to consolidate bonds between members to respond collectively to the crisis. In places where both types of crises exist—a disaster event and violent conflict—this difference in viewpoint about each type of crisis can be used creatively to further community resilience through using discussions and consensus building on how to deal with natural disasters that may then build social links and agreed strategies that can be used by the community to respond to the conflict as well.

References

Ahmed, R., Seedat, M., van Niekerk, A., and Bulbulia, S. (2004), "Discerning Community Resilience in Disadvantaged Communities in the Context of Violence and Injury Prevention," *South African Journal of Psychology*, 34: 386–408.

Anderson, M. B. (1999), *Do No Harm: How Aid Can Support Peace—Or War*, Boulder: Lynne Reinner Publishers.

Anderson, M. B. and Woodrow, J. P. (1998), *Rising from the Ashes: Development Strategies in Times of Disaster*, Boulder: Lynne Rienner Publishers.

Balvin, N. (2007), *Understanding and Engagement in Peace Building: A Review of Emerging Issues and Related Activities*, unpublished report for UNICEF headquarters.

Battro, A. M. (1973), *Piaget: Dictionary of Terms*, translated and edited by Ruttschi-Herrman, E. and Campbell, S., New York, Toronto, Oxford, Sydney, Brausschwsig: Pergamom Press.

Boulding, E. (1990), *Building a Global Civic Culture: Education for An Interdependent World*, Syracuse, NY: Syracuse University Press.

Boulding, E. (2000), *Cultures of Peace: The Hidden Side of History*, Syracuse, NY: Syracuse University Press.

Bronfenbrenner, U. (1979), *The Ecology of Human Development: Experiments by Nature and Design*, Cambridge, MA: Harvard University Press.

Brown, D. and Kulig, J. (1996/97), "The Concept of Resiliency: Theoretical Lessons from Community Research," *Health and Canadian Society*, 4: 29–52.

Cessaro, M. A. and Romero, E. M. (eds.) (1987), *The Mexico Earthquakes 1985: Factors Involved and Lessons Learned*, New York: American Society of Civil Engineers.

Coleman, J. C. (1988), "Social Capital in the Creation of Human Capital," *American Journal of Sociology*, 94: 95–120.

Coles, E. and Buckle, P. (2004), "Developing Community Resilience as a Foundation for Effective Disaster Recovery," *The Australian Journal of Emergency Management*, 19: 6–15.

Conley Tyler, M., Bretherton, D., Halafoff, A. and Nietsche, Y. (2008), "Developing a Peace Education Curriculum for Vietnamese Primary Schools: A Case Study of Participatory Action Research in Cross-Cultural Design," *Journal of Research in International Education*, Vol. 7, No. 3: 346–368.

Folke, C., Carpenter, S., Elmquist, T., Gunderson, L., Holling, C., Walker, B., Bengtsson, G., Berkes, F., Colding, J., Danell, K., Falkenmark, M., Gordon, L., Kasperson, R., Kautsky, N., Kinsig, A., Levin, S., Maler, K., Moberg, F., Ohlsson, L., Ostrom, E., Reid, W., Rockstrom, J., Savineji, H., and Svedin, U. (2008), "Resilience and Sustainable Development: Building Adaptive Capacity in a World of Transformations." Online. Available at "http://www.resalliance.org/576.php" www.resalliance.org /576.php, accessed October 31, 2008.

Galtung, J. (1978), "Peace and Social Structure," *Essays in Peace Research Volume 3*, Copenhagen: Ejlers.

Inter Agency Standing Committee (IASC) Policy Statement for the integration of a gender perspective in humanitarian assistance, Geneva May 31, 1999. Online. Available at http://www.icva.ch/doc00000755.html, accessed April 20, 2011.

Jabri, V. (1996), *Discourses on Violence: Conflict Analysis Reconsidered*, Manchester: Manchester University Press.

Lederach, J. P. (1997), *Building Peace*, Washington, DC: United States Institute of Peace.

Lederach, J. P. (2005), *The Moral Imagination: The Art and Soul of Peace Building*, New York: Oxford University Press.

Norris, F. H, Stevens, S. P, Pfeffererbaum B., Wyche K. F., and Pfefferbaum R. L. (2008), "Community Resilience as a Metaphor, Theory, Set of Capacities, and Strategy for Disaster Readiness," *American Journal of Community Psychology*, Dartmouth Medical School: Hanover, NH, USA.

Organization for Economic Cooperation and Development (OECD) (2001), *The Wellbeing of Nations: The Role of Human and Social Capital, Education and Skills*, Paris: OECD Centre for Educational Research and Innovation.

Paton, D. and Johnston, D. (2001), "Disasters and Communities: Vulnerability, Resistance, Resilience and Prepardness," *Disaster Prevention and Management*, 10: 270–277.

Pfefferbaum, B., Reissman, D., Pfefferbaum, R., Klomp, R., and Gurwitch, R. (2005), " Building Resilience to Mass Trauma Events" in *Handbook on Injury and Violence Prevention Interventions*, New York: Kluwer Academic Publishers.

Piaget, J. and Inhelder, B. (1969), *The Psychology of the Child*, New York: Basic Books.

Poniatowska, E. (1988), *Nothing, Nobody: The Voices of the Mexico City Earthquake*, Philadelphia: Temple University Press.

Putnam, Robert (1993), *Making Democracy Work: Civic Traditions in Modern Italy*, Princeton, NJ: Princeton University Press.

Putnam, Robert (1996), *Bowling Alone: The Collapse and Revival of American Community*, New York: Simon and Schuster.

Resilience Alliance. Online. Available at "http://www.resalliance.org/576.php" www.resalliance.org, accessed October 31, 2008.

Savitch, H. V. (2008), *Cities in a Time of Terror: Space, Territory and Local Resilience*, New York: ME Sharpe Publishers.

Shamai, M. and Kimhi, S. (2004), "Community Resilience and the Impact of Stress: Adult Response to Israel's Withdrawal from Lebanon," *Journal of Community Psychology*, 32: 439–451.

Sonn, C. and Fisher, A. (1998), "Sense of Community: Community Resilient Responses to Oppression and Change," *Journal of Community Psychology*, 26: 457–472.

Sugarman, S. (1987), *Piagets's Construction of the Child's Reality*, New York, New Rochelle, Melbourne, Sydney: Cambridge University Press.

Treaster, J. B., "Mental Distress Stalks Mexicans," *New York Times*, September 27, 1985, p. 8.

United Nations Development Programme (2004), *A Global Report: Reducing Disaster Risk, a Challenge for Development*, New York, NY: The Bureau of Crisis Prevention and Recovery.

Walhstrom, Mary (2007), "Foreword" in *Building Disaster Resilient Communities: Good Practices and Lessons Learnt*, Geneva: UNDP, p. iii.

2

INDONESIA

Dicky Pelupessy, Diane Bretherton, and Anouk Ride

Dicky Pelupessy holds a bachelors degree in psychology from the University of Indonesia and a masters degree in state management and humanitarian affairs from the University of Rome "La Sapienza," Italy. He has a passion for humanitarian work and has been working in crisis and disaster-affected areas in Indonesia since 2001. He was the coordinator of the Aceh area of the Pulih Foundation, a national NGO working for trauma recovery and psychosocial intervention. He is a lecturer in the Faculty of Psychology at the University of Indonesia and now the director of the Crisis Center of the Faculty of Psychology, University of Indonesia. He is also the chairman of the executive committee of the Indonesian Network of Psychosocial and Mental Health for Disaster Management.

In my mind I said: "Ya Allah, today I start my life from zero. Ya Allah, today I become alone again." And then suddenly in my mind again: "No. you ar not alone. You are not starting from zero." And then I said: "Yes, I'm not alone. I'm not starting from zero." I said: "I have knowledge. I have friends that make me strong." But as strong as I was, I also cried. I cried anytime I go while calling the names of my children. I saw the dead bodies there. It was something like everything happened.

> Tabrani Yunis, director of the local Centre for
> Community Development and Education (CCDE)

People get a lot of advantages from the tsunami. One of them is they could feel thankful. Yes, thankful 'cos it was a disaster. It's not just us who got it but everyone. We were probably too proud with what we had, then suddenly the disaster came and everything was gone. And we couldn't say, "oh don't take my house, don't take my children,"

I heard mothers, older women talking about that. So, they became tougher, they became more... "oh it's not just me but everyone."

They could be together. So, when problems occurred, like when we were at barrack [camps]: "If we're removed and got no place to stay, we have to stand together. Every one of us should get it," and not only at barrack, also at the village. Like me, I didn't just live at barrack but I got a village too, at Uleelheu. So, if one of us didn't get a house, everyone wouldn't accept it. That's the way it went. So, we got each other's back.

Novi Kurniawati, University student and NGO worker

CONTEXT

Since it lies on the edge of the Pacific, Eurasian, and Australian tectonic plates, and has 150 active volcanoes, Indonesia is a common site of volcanic activity and earthquakes—the most famous being Krakatoa's volcanic eruption, which erupted in 1883.

More recently, Aceh, an area of mountainous forests and coastal environments located at the westernmost tip of the island of Sumatra in the north of the country, became the site of Indonesia's most dramatic natural disaster—the 2004 tsunami. The affected capital, Banda Aceh, is home to 250,000 people, and approximately another 4 million people live in other towns and rural areas in Aceh (the overall population of Aceh is 4,271,000).

As the largest archipelagic state in the world, Indonesia has islands of forests, mountains, and volcanoes, ringed by coasts of mudflats, mangroves, and coral reefs. With 26 percent of the 1,531 species of birds and 39 percent of its 515 species of mammals being unique to Indonesia, it has high biodiversity despite significant human changes to the environment—mostly due to wide-scale logging for plantations and urban expansion.

Heading such economic development in Indonesia is an elite group that has attained wealth through links to political power. Access to government, including appointments particularly in the era of President Suharto, has enabled the elite to establish their control over exploitation of natural resources and government services. Transmigration and settlement programs have also put people from the center of Indonesia, Java Island, out onto the land in the provinces, fueling disputes over land, land use, religion, and conflicts between different cultural and ethnic groups.

Major industries such as logging, paper and pulp production, and palm oil were the result of these programs. Today, major exports of Indonesia include oil, natural gas, crude palm oil, coal, appliances, textiles, and rubber.

Aceh's trade was originally centered on the spice trade. In the 1820s, it was the producer of over half of the world's supply of black pepper and a key part of the Dutch East India Company operations in Asia. After the independence of Indonesia in 1945, the primary economic development in Aceh was oil and gas from the Arun natural gas field. The extraction, managed by Exxon Mobil, became controversial—in the first instance because of the dominance of taxes and other local economic benefits going to Jakarta rather than to the province, and in the second instance because villagers sued the company for its role in supporting equipment and infrastructure for local army operations against rebel Aceh forces. Despite the controversial nature of this development, 30 percent of Aceh's GDP was from oil and gas mining in 2004 and 30 percent of GDP was from agriculture and fisheries (which supports livelihoods of the majority of the people) (Nazara and Resosudarmo, 2007: 3).

Indonesia ranks 111th on the United Nations Development Programme's (UNDP) Human Development Index—above Vietnam but significantly below other Southeast Asian countries such as Thailand at 87th and Malaysia at 66th (UNDP, 2009: 168). It has a relatively high life expectancy of 70.5 years and a mortality rate of children under five of 36 per 1,000 (UNDP, 2008: 263).

While Aceh, compared to other provinces of Indonesia, had a relatively large per capita spending on health and education, it also had a higher poverty rate of 28.4 percent compared to 16.7 percent in Indonesia generally in 2004. In Aceh, the poverty rate in rural areas (32.6 percent) was almost double the rate in urban areas (17.6 percent), reflecting the differences in education and income between the capital Banda Aceh and the rural areas (World Bank, 2008: 13). In Aceh, being well-off means owning houses and recent models of cars. Some Acehnese with education and resources have relocated to other parts of Indonesia and abroad, particularly during periods of increased violent conflict.

Aceh, a remote Muslim province with a separatist rebel movement, martial law, and government-restricted access to outsiders, had less aid than other parts of Indonesia before the tsunami, many places experiencing outside assistance for the first time after the disaster. Indonesia experienced a dramatic rise in aid after the tsunami, with just $0.4 per capita in 2004 (UNDP, 2006: 345) rising to $11.4 per capita in 2005 (UNDP, 2007: 292). The number of national and international NGOs in Aceh rose from just 12 before the tsunami to 300 after the tsunami (Brusset,Bhatt, Bjornestad et al., 2008: 36).

The arrival of aid contrasted with the regional identity of Aceh as independent and distinct from the rest of Indonesia. The country's 220 million people, spread across 10,000 islands, encompass many different cultural backgrounds and languages, with Bahasa being a language created to help national unity.

Of all the territories of Indonesia, Aceh is perceived as the most adamant to continue its traditions of intermingling politics and religion. Originally a sultanate, Aceh reached the peak of its powers under Sultan Iskandar Muda (1607–36), who waged a war against the Portuguese to get them out of the Straits of Malacca that ultimately failed. Aceh sided with England to fend off the Dutch, and a trade agreement with London has been used as evidence of an early Acehnese claim to statehood. The 500-year independence of the sultanate finally came to an end after 32 years spent fighting the Dutch.

After the Second World War, Aceh joined the new independent Republic of Indonesia, but despite being a province with increased autonomy since 1963, Acehnese political leaders quickly became disillusioned with the central government, which they saw as corrupt, exploitative, and un-Islamic. Economic exploitation of Aceh's fossil fuels and forests, coupled with transmigration policies, became grievances for the Acehnese.

In 1977, rebel group Gerakan Aceh Merdeka (Free Aceh Movement or GAM), led by local entrepreneur Hasan Di Tiro, proclaimed independence—though Di Tiro was later forced to flee to Sweden. Fighting continued to oppose the regime, then President Suharto's fall and a flourishing of freedom of speech and democracy in the late 1990s also spread to Aceh and gave the Acehnese the ability to begin a peaceful movement for a referendum. As in other parts of Indonesia, these pro-democracy actions were dominated by students and were demonstrated through banners and public meetings.

However, the conflict between the army and the rebels continued and the International Organization for Migration estimated that 320,000–350,000 people, out of a total population of 4 million in 2004, had experienced displacement due to intensified conflict and violence in Aceh since the fall of Suharto in May 1998. One survey of 1,700 Acehnese indicated that a majority of Acehnese had viewed or experienced conflict, 53 percent of people surveyed were forced to flee violence, and 39 percent knew a family member or friend killed in the conflict (Grayman, Delvecchio, and Good, 2009: 295). Others had experiences such as being forced to search for or identify family members (28 percent) and being publicly humiliated (13 percent) by military or rebel forces.

Culturally, the Acehnese share some outlooks and traditions with the Malays and the Arabs, which have influenced the development of the Acehnese language. The *meunasah* (mosque) in Acehnese communities is a community gathering point that is central to religious education and community activities and services. Without access to many international organizations, the Acehnese have long established and run local NGOs such as religious groups, community-based economic enterprises (saving clubs, village development associations), and social groups (such as for organizing funerals).

This daily life was disrupted at 7.58 A.M., on December 26, 2004, when an earthquake measuring 9 on the Richter scale, struck off the coast of Sumatra, triggering the tsunami with its waves of up to 30 meters that would affect coastlines in ten countries. However, the most affected population was in Indonesia, with 500 kilometers of coast destroyed, 130,000 people dead, and an estimated 500,000 displaced.

In Aceh, 116,000 houses were destroyed and 12 percent of the population was displaced by the tsunami. Survivors camped in mosques, schools, stadiums, and open fields. The mortality rate of women was four times that of men, leaving many children in the care of extended families and an increased burden on women survivors (Carballo and Heal, 2005: 12). Over 100,000 small businesses were destroyed, and more than 60,000 farmers were displaced due to damage to land and property.

Many people lost or had their homes and their livelihoods (small businesses, farms, fishing) severely damaged. The tsunami damage (estimated replacement costs of infrastructure and lost income) was 4.4 percent of the Indonesian GDP. Among India, Indonesia, and Sri Lanka, the proportion of losses in income was greatest in Indonesia, where 83 percent of those affected lost more than 50 percent of their income (Fritz Institute, 2005: 4).

RESEARCH OUTLINE

Affected people were asked about their experiences through 12 interviews conducted in Aceh in 2009 with people who had first-hand experience of the tsunami (from those whose experience included being swept away by the water to those who had been in Aceh at the time of the disaster) or had played an active role in the community responses. These included volunteers and humanitarian workers and community members who participated in NGO programs with survivors who were actively involved in organizing fellow survivors. The interviewer, Dicky Pelupessy, had experience working on psychosocial support programs in Aceh, and many of the interviewees were personal contacts who had also been involved in responding to disasters. All of the interviewees spoke Indonesian so no interpreters were used, but the interviews were later transcribed into English.

COMMUNITY RESPONSES

Since it was a Sunday morning, many people were at home or heading down to the sea when earthquakes struck the low-lying capital of Banda Aceh and the surrounding areas. People had been through earthquakes before, so only became concerned when it seemed to have more force than normal. Then the earthquake was followed by waves of up to 30 meters, creating confusion about where and how to escape. No one received any official advice or warnings; instead, people relied on the knowledge of the people around them, none of whom had experienced a tsunami before.

Some had heard of Aceh's experience of a giant wave in the past. Daudy Sukma, a student and an activist from Aliansi Beudoh Aneuk Nanggroe (abbreviated ABANG), a local NGO focusing on child and adolescence issues, said,

> All I know, is what my parents usually told me in elementary school: *Ibeuna*. It was water that reaches the sky. It's either a legend or else, the point is that is what my parents told me. It depends on which side the water falls. If it falls to that direction, if it falls to our direction, then we will be wiped out. If it falls to the left, then the left side will be wiped out. So it's said, this thing happened before, but I didn't know. Didn't know what a tsunami is and what makes a tsunami.

When the earthquakes became more severe, people began to act, as Edy Syahputra, a law student, working with Anak Bangsa Foundation on street children programs, described the experience: there was panic in his coffee shop when the first earthquake hit, with people falling, electric poles shaking, and sounds like explosions. Then, things became even more worrying and baffling:

> The second earthquake came, and more terrifying than before. There was a wall in my house, the wall was high, when the second earthquake came, the wall fell down on houses around me. When the wall was falling, I spontaneously try to get my family. Thank God, they're okay; there were my two little sisters, my little brother, and my mother, four

of them. I got them okay. I asked them to go to my shop. While sitting there, I heard news, I don't know where it came from? Maybe after the first earthquake, people went somewhere? I heard that a big building fell down in the city. Investigation found that it was Pante Pirak. Pante Pirak had fallen. People crowded around sitting in my shop. People [were] already panicking. While sitting there, there was someone that said: "Praise to God." In a second, I saw many people crowded from Lampulo, screaming and panicking terrifyingly and I really didn't know what it was. But the information was really fast: "Sea water came up." I thought when people said that sea water came up to the plain it was so bizarre. I still did not believe that sea water would come up to plain. And, behind my shop, there was a path, this was my shop, this was a path, there are so many houses. I saw that the reaction was the same there. This line, behind it, there was a separation embankment, that was Aceh River. And people panicked. A terrifying panic. Screaming "Sea water came up! Sea water came up! Sea water came up!" that was the only sound that I heard from them. They were screaming many times. I could not imagine how to imagine...I saw the waters turned from green to black slowly. It turned slowly, not quickly like bam! It turned slowly. Everyone was running, not even thinking to climb the buildings.

People close to the sea realized more quickly that something strange was happening, as Rumiati, a local housewife, described that she noticed the sea dry for about two kilometers out:

"Oh why is the sea dry like this? I've been living here for 14 years," I wondered. "the sea has never been this dry. What's happening?" I called my neighbor. "Wi," I said. "come here. Look, the sea's dry," "why's it so, *mbak* [woman]?" "I don't know. I've been living here for 14 years and the sea has never been like this." Because I lived there first compared to those people. They're new.

Then I...I don't know why but my heart said, "oh let's just go up there." Yes. I never heard before that there'd be an earthquake, tsunami, or rising sea, I never heard of them.

"Maybe the sea is parting," so I said to my neighbor I was talking to before, "I'm afraid it will go up really high." So I told her to go up to the higher place with me.

The waves were unlike anything people had ever seen before. Daudy, student and activist, remembered his thoughts at the time:

Oh, that isn't like water...that is like an animal or something, a creature that is very tall, stands sky-high and can destroy everything. Personally, I imagine it as the most horrifying thing in my life.

The concept of a tsunami was unknown to many people, and those who knew about it thought it occurred in other parts of the world. Edy, street children worker, said few people had even the vocabulary to describe the disaster:

> This was the first time I heard the word "tsunami." People said, in Aceh, it was called *Ibeuna*. Also it's the first time I heard this word *Ibeuna*. I saw on TV that tsunamis often come in Japan.

Ahyar Rasyidi, a community worker, said at first people thought that people were running because there was a fire, and cried: "War!" Erni Munir, a midwife, said people panicked, ran, and cried out to God. She told her children: "Don't scream, praise Allah. We got calamity, so praise Allah."

Some people could not react during and after the disaster—Novi, student and NGO worker, describes herself standing like a statue in shock, some people were crying out and not moving and needed to be prodded into action by other people. The loss and fatalities were so great that many people could not believe what had happened.

Edy, the street children worker, recalled the horror of walking around after the waves stopped and realizing that there were so many dead and grieving:

> There is 20 meters to get from that street to the house. Corpses splendid around the street. I didn't know how to explain those awful forms. Naked, headless, and other awful things that I have never seen before. There were corpses buried in soil. The streets weren't flooded but the village was. Panglima Polem is still in the city but there were houses in the village. Citizen's houses were still flooded and I saw dead bodies everywhere.

The immediate concern of everyone was to try and find family members and discover if they were safe or to retrieve their corpses. Amid the grief and crying, many people told distressed people to focus on the practicalities of saving themselves and finding their family members.

People described the day as an unimaginable experience—fish and sea foam were in rice fields, volleyball fields had cracked in half, cars had been washed away, corpses were in trees, and people were trying to find their family members as the city and towns had no electricity and communications, and little working transport. Asrida Vonna,

"Ida," a volunteer at Pulih Foundation, said that when she was running she thought it was the end of Aceh:

> There was a country, in Europe, isn't it? Mesopotamia, which had sunk in the bottom of the sea. So, I thought, maybe in a few moments...Banda Aceh might be sunk under the sea. When I was running, I was thinking..."it's doomsday."

People tried to get to higher ground, sharing cars and buses or running on foot. Some people went to the roofs of houses, barns, and shops to escape the water, others to hills. Novi, NGO worker, describes running to a car and the kind nature of the driver who helped her:

> I got into the car at the front. Lots of men were already sitting at the back, it was crowded. I got in the car and the driver was saying prayers. "Subhanallah, ya Allah, is this doomsday?" he said. He spoke to me in Aceh language: "Relax." He asked me to relax and not to panic. I said, "No, sir, I don't." He just kept saying prayers. "This seems to be doomsday, no more us," he said. Then the car moved slowly because there were so many cars in front of us. The distance was about 500 meters, we were far apart from the car in front of us. It was all water behind us.

> Then suddenly the driver stopped; in front of Perumnas 1, before we reached Brimob. He stopped all of a sudden. The assistant asked: "Why stop? Don't stop, sir. Just keep moving." "It's okay, there are angels," the driver said, "we're waiting for those angels." I was curious: "Which angels, sir?" He answered: "There, little kids." It turned out to be kids running from the water, one of them was 6 years old, the other one was his brother. The driver stopped and turned the engine off. His assistant said: "Don't turn off. Just let them run and catch us, we just keep moving slowly." The driver answered in Aceh language: "Love the angels." After they're close enough, the driver opened the door. The door on his side: "Hang there. Keep hanging on the door," he said. They hung their feet, and *Subhanallah* [thanks Allah], a huge wave came, less than 10 meters in front of us. It was all in front of us, I couldn't say anything.

> We weren't struck by the wave. The huge wave was exactly in front of us, taller than the coconut tree. It was...in front of my eyes. The driver was saying prayers, *Adzan*. The kids were climbing to the car, the engine hadn't been turned on by the driver. Then the huge wave. I couldn't say anything. The wave rolled up thousands of people in front of us. The traffic was stuck, it was crowded. It was about 500 meters from our car, exactly in front of us. The driver said, he snapped at me and told me not be shocked like that. He asked me to get off the

car. "Kid, kid, kid, get off the car." I answered, "Where?" "Anywhere people run." "How can I run?" I said. The water was behind me, beside me, and in front of me. "I don't know where to run!" so I said. "Just run anywhere," he said. "Go! Run!" and those two kids were nowhere to be seen.

Ahyar, community worker, recalled being caught up in the water and the people in trees and on the roofs escaping the water:

> When I turned back, turned out it was even bigger from behind, because it was from the sea. So, I stop there in the middle. In the middle, rolling around and around in the rushing flood, for a few seconds, or maybe a minute and then I got saved because the water pushed me to the middle of the storehouse, there was an empty space, if I got slammed into the storehouse, I might have died, but I didn't, suddenly I was passing through a tree, so I caught the branch, and at the tree's body, there were lots of lizards. And there was a dog also, and so I stood there. At first the water rose and kept going up and up, and just like in the movie. It stopped at the nose level. I think, it could be bad if the water kept rising, but it didn't, and so I stood there, and the rushing water came for three more times, for about half an hour, in front of me, there's this lady in the roof. She said: "Come up here!" And I replied: "No...I'd rather stay here." Because it could be dangerous if the earthquake came, so I was safer on the tree. And after half an hour, after no signs of rushing water anymore, I climbed down the tree. And went up the house and when I was on the roof, earthquake came a few times, so I came down.
>
> There was something funny that time—in Darussalam there was this crazy person, named Alu. The crazy thing about him is that he walked around in *sajadah* [a prayer mat]. He was the first one to climb down to the water, when nobody else dared to do the same. "Oooo...Alu is all right..!" they said. And, so, they all dared to climb down. And I climbed down to the restaurant's fence, because it was high and there were people above, on the roof. How did they climb up there, I wondered? But anyway, lots of people on the roof. That was what happened.

Neighbors and families held together at this time to try and keep safe, and there were also many people who helped strangers who were troubled, in the same situation as them or in worse situations. In each group of survivors there were people who identified solutions to problems and helped those who were too afraid or too shocked to respond to the situation. For example, Novi, NGO worker, described a bunch of people on the roof of a store in Simpang Tiga—included in this group was a person who told the women not to cry but to pray, another person who led efforts to break the roof to get people to higher ground, and a group of men who went into the flooded regions to try and rescue people after the first wave. The second wave killed some of the men who had gone down to do rescues, and Novi describes a man leading the singing of the *Adzan* [prayers] to try and calm the distressed people watching from the roof as the men doing rescues were washed away by the second wave.

People cooperated to get and share food and clothing (the tsunami striking in the morning meant that many people were still sleeping and were either naked or half dressed). Novi, NGO worker, described:

> There were women and girls who were naked. They were given those clothes [from the shop], and some gave their own clothes. Then the men went down, they took food; noodles in plastics, in the water, they pulled them out. Mineral water, fruit tea, and anything in cans, they took them all and threw them up. We all had to drink. "Its okay, the shop keeper had given the permission. Just take them all."

Many stories were told of people cooperating to get and share food and of local businesses supplying food and water:

> By afternoon [of the day of the tsunami] someone was already giving out food. I saw people given food by a man, who had come there by car. Some people came by train, and they're giving out mineral water too. We didn't know where to go so we stayed under the roof of the shop. A woman said: "Come in! It's a lot of us here, from the 1st to the 3rd floor. Just come in and drink here." She gave us drink and bread, perhaps because she saw us looking pale.... Then there was another man, he seemed rich. He came to give out rice and water for us who were sitting and crying. "Eat this, kid. Keep this bread and eat later." (Novi, NGO worker)

> The positives that I've seen after the tsunami, there's some people who were stingy, after tsunami, but others where everything that is in her/his house is taken out and given to other people. There's a midwife too, whose house has become like a barrack (camp). Like a barrack, she kept giving until she couldn't remember how many clothes that were in her wardrobe, were taken out. That's the positive side. (Erni, midwife)

The anxiety, fear, and grief was great following the disaster. Daudy, student and activist, described 300 people sleeping outside on the ground, as people were too afraid of more earthquakes to go inside. Ahyar, community worker, said,

> For a month, Aceh wasn't really a comfortable place. Because of that, many went out of Aceh. When I saw their faces...they all showed the same thing...they didn't want to stay here anymore, because, well, if you look at the writings on destroyed buildings, it said, "Goodbye Aceh" and sort of expressed that people were frustrated. It was a terrible month.

However, many people stayed and started efforts to help themselves and others affected by the tsunami. Ahyar, community worker, explained her choice:

> I decided to stay in Banda. First of all, it is my home. Second, I have a lot of friends there. Many of my friends were making post [assistance], and I helped them, like carrying the dead bodies, distributing goods, and I didn't worry about food, because lots of friends were there. So, I decided not to go, well, while I could help the society. Even that time, a month after the tsunami, help came from outside...there was not so much help on the first or second weeks.

Dian Marina, Aceh area coordinator of Pulih Foundation, a local NGO working for trauma recovery and psychosocial intervention, said people tried to cheer one another and help with the range of reactions that people had to the disaster,

> Some people said: "Let it be, it's Allah's decision, we have to be patient, to admit it," "there's a wisdom behind all of this" like that, right? "Be patient." There's different reactions, there are people who fainted, there are people crying, there are confused, there are quiet ones with an empty gaze, there are some going crazy.

After saving themselves, locating their families and belongings, people began to form groups, said Dian, NGO coordinator. Those who were coping well with the situation became leaders at this time and organized activities such as public kitchens.

Tabrani Yunis, director of CCDE, said people's volunteerism and action was a way of dealing with personal troubles,

> So that's why I contacted friends from Jakarta to send me some money. I wanted to help raise the children. So, I purchased some school uniforms and then I distributed by myself in Southwest Aceh and then

more friends give me money, and then I bought more uniforms until I could help more than 500 children. I did this even though I was in sad. I think I decided to help because I realized this was my work. And also this was one way to make me strong. If I just sat down and was thinking, maybe I would be crazy, my mind would go somewhere... what happened to me... I would be thinking and cry and cry. So maybe this is one solution for me to make me, yes, secure.

Village *keuchik* (village head) or imams and village organizations had a key role in cleanup efforts in many cases. Asrida, volunteer, said there was a sort of "togetherness" at this time with custom factors influencing togetherness: "Like the *keuchik's* role, the role of the *ulamas* (religious leaders) to give the spirit back to the people."

Fatmawati Luthan, "Fatma," former office manager of Yayasan Anak Bangsa (Anak Bangsa Foundation), a local NGO working for child protection issues, said religious activities, cooking for women, building houses for men, and young people helping these activities in the barracks were under way:

> So, those who didn't use to have any activity, then did some activities with the others. Finally, we as victims tried to make each other stronger. Even though I was a victim, I didn't want to put myself in that position. I had to get out of that position, I had to make myself stronger.

Dian, NGO coordinator, said people turned to religion after the tsunami:

> At that time, I remembered in every mosque, there are always religious lectures, and the mosque is filled after the tsunami. Maybe because people felt sinful ya? It was so full because of people listening to religious lectures, and it gave us peace.

However, people also wanted to blame others for the disaster, and some people said it was a punishment for immorality. Edy, street children worker, explained the thinking of some people:

> People argue that the tsunami was caused by sins we made. Because we did evil things that were prohibited by religion, we were cursed by Allah. If we didn't want to re-experience the disaster, we had to repent and ask Allah for forgiveness. Well, what things did we do? I'm confused looking for the connection. For example, if we go dating, we

will hold hands spontaneously. They said it is a sin. We are prohibited to do that again. I think it is strange, connecting two things that are not correlated. Until now, I still don't understand the logic.

Others saw their survival as a positive sign from Allah who had spared some people so that they could live better lives. Erni, midwife, described what led her back to work with women as a midwife:

Something came up on my mind, some desires, that maybe, God saved me because I'm still needed here. That's why my faith grows stronger and I become more loyal to Allah. If Allah has not wanted to take us away, there are reasons, there's a way to be saved. Just like me, I was under the water, messed up, but Allah said, "I don't want to take your soul away right now." That's what made me come to the barracks. I conclude that I have to do my job. So on the 44th day of tsunami, I practice on the road. At Leungbata road. I went out every day with bandages. My first purpose, was to find the dead bodies, I had 3 daughters. After that, we were at Neusu, it was so crowded, it might be, one house for 50 people. So, I felt that there's a moral responsibility that those people have to eat. So, I went to the Parliament office, and there was someone who gave hulled rice. Well fortunately, that person is my patient, so I took 2 sacks to go home.

Fatma, NGO manager, described people's need for solidarity and listening at that time, which was helped by local NGO visits to camps:

At a barrack in Lambaro there was a woman who always cried every time she saw me and Evi. If she saw me and Evi, "Fatma, teach me first so I can stop crying." "It's okay, ma'am. Just cry if you feel sad." But at last she felt happy that there's someone visiting her, asking about her condition. Even though it was hard for her to say it, but she was relieved after telling me and Evi. Women there are all happy if they're visited.

People who had taken in those that were left homeless by the tsunami became creative about how to provide food for everyone, as described by Nur Janah Nitura, "Nurjanah," a longtime practicing psychologist in Aceh, whose house became a warehouse for supplies and whose car was rented to foreign journalists to pay for food for the refugees.

She describes teams of local psychologists and volunteers who visited barracks to talk to the women refugees and try and assist:

The schedule was up to us, nothing fixed. We got together in the morning: "Where are we going today?" "Today we go to . . . ." "What do we do there?" . . . "We'll go to this place, social building." "What do we do there?" "Well just give massage if you can, or listen to the stories if you can. There's no such thing as theories now, they just need a place to share." We kept listening until they felt relieved. Then we'd give them massage, (commonly done in the house) *"Astaghfirullahaladzim..."* [I ask Allah for forgiveness] they'd say. It means they felt better, someone had listened to them so they felt quite relieved. Then they'd ask: "When are you coming back here?" "Come back, okay?" "Okay." Then the team shared stories on the way back in the car. "What did they do?" "Oh they were bla bla bla..." we told each other this and that, one said: "Oh I was close to getting driven out!" Then we'd laugh together.

Even though many people had lost their homes and livelihoods, community achievements included finding family members and corpses, finding and sharing food and clothing, plus psychosocial support and youth programs. While the scale of international assistance was large, much of it did not arrive until a few weeks after the tsunami and was spread haphazardly, meaning the initial response and recovery efforts were largely local and community based. For instance, a Fritz Institute Report (2005) indicated 91 percent of the rescue services were provided by local individuals, and that of all tsunami-affected countries Indonesia was least satisfied with the assistance provided by others within the first 48 hours.

Assistance for people affected by the tsunami was provided by religious groups, the government and army, local NGOs and international NGOs, and included food, health services, basic materials (e.g., cooking materials and clothing), and housing as well as a range of social supports and programs such as counseling and arts programs.

In the first instance, the most common way people sought assistance was to go to the mosque. Many interviewees described the mosque as a place people came to after a disaster or dispute and a place where information is exchanged through sermons, schools, and social gatherings. Mosques also distributed assistance such as supplies and arranged burials.

As mentioned above, the tsunami assistance was the first experience most Acehnese had with international organizations and other foreign personnel (such as foreign soldiers who assisted with collecting and identifying corpses). People in Aceh described the positive effects of outside aid and assistance in practical and spiritual terms. Edy, the street children worker, said the effects of assistance were such that

we started to breathe again. We all were destroyed but then the help was coming. Cooking utensils, camps, foods, and anything, the village which had been broken, started to come back.... The village was still silent, dark, and quiet but people decided to live in it. To settle, hold out, start activities, start a new life.

For Dian, NGO coordinator, the fact that people from outside were sympathetic was encouraging:

> What do we live for when you have nothing—no family? Then when they receive help from others they finally think that many other people still care about me. So that it helped them to stand up and continue their life. Many people from many skin colors, religion, that Acehnese were resistant to before: "Who's that *kafir* [non believer]?" At least when they see what happened, they know that they're not as bad as they think. That we are family. We are all family and we have to help each other.

Many people affected by the tsunami stayed in their village, often sleeping outside in groups while housing was rebuilt. In villages, communities often distributed assistance themselves as Edy, the street children worker, explains:

> We didn't choose anyone as a chief but we connected with the *lurah* [leader] in that place. We asked *lurah* for information about people who needed help.

Other people sought shelter in camps. Internally displaced persons were grouped into camps, first in tents, then in barracks (barrack refers to a more solid shelter, made from plywood and wood, but still temporary), some of which were organized by the Indonesian government and army. People came from different areas to these camps, which sometimes had poor living conditions that added stress to the lives of people affected by the tsunami.

Erni, midwife, describes her discomfort at the lack of dignity accorded to the people at a distribution center:

> There's an incident that made me sad. At the beginning, I forgot on what day, before ten days, in front of Pante Pirak, there's someone who gives hulled rice, oil, sugar, etc. So, that time, I was wearing the dress that I told you, for a week. And then, when I came, I felt something, I go to the front row, "Sir, I want to...." "Hey! You want hulled rice?! On the line!" he said. I was so devastated "God...this is how it feels to be poor people" that's what I remembered. I cried that day. "If you want it, on the line!" they said...*Astagfirullahaladzhim*...[I ask Allah for forgiveness] I cried there. I cried and waited.

Instead, Erni had 50 people living in one house, groups of which went out each day looking for food they found floating in the waters until donations for her came six days after the tsunami.

Barracks were described as difficult situations for children and women—there was no privacy for families and women felt unsafe, particularly while bathing. Novi, NGO worker, described her volunteer work at the hospital as a break from difficult conditions at the camp including sickness, arguments, feeling restricted (because she could not go out alone with so many men out there), and unsettled about different people living together and clashing over different styles of communication and behavior.

The Indonesian army had a role in responding to the disaster in dealing with debris and the organization of refugees and their security, but in the initial instance in several cases people said that the army appeared unprepared and unresponsive to the people. One interviewee said that the soldiers around a hospital acted inappropriately:

> At the hospital, I met my neighbors. Oh, they were safe. And I kept hoping that my family was safe too. I spent one hour there and realized that just few nurses worked. A lot of soldiers just had fun all around and laughed without knowing there were a lot of people suffering from disaster. I thought they acted like animals.

Another interviewee said that there was unease among local people about the presence and activities of military personnel:

> Soldiers were plentiful in those early weeks. From Australia, America, even Red Cross. Probably because you couldn't come in [to Aceh] that time, the military were still around. There was this American soldier who hit a kid. We said: "You can't do that," and so we didn't accept that soldier. The people didn't really fit with the army. There are stories that we heard in Calang, that the foods were piled up in the army's barrack, not the people's. Well they probably thought that all of the people in Aceh were GAM (rebel forces), so just let these people die, ha ha.

People described asking the soldiers for assistance to carry dead bodies and being turned down.

Some local people corralled together those who wanted to help— Erni, midwife, mentioned staff from Jakarta from UNICEF, WHO, and Save the Children who helped her and other health workers provide maternal health services to women with assistance from Johnson and Johnson, who funded the provision of vouchers for free checkups for one year. Other organizations recruited local volunteers; many "paid volunteers" to join activities. Novi, NGO worker, described being approached by a UNDP coordinator who suggested that she come and volunteer to help at the Zainal Abidin hospital:

> Some cleaned up the hospital, some cleaned up the yard, some cleaned up the rooms and some helped the patients. I helped the patients, the children. "Oh I want to help, I don't want to stay here anyway," I said. So we were picked up at 9 and came back at 6, we were taken back to the reception centre. It lasted for 4 months. And they gave us clothes.

Volunteering, said Novi, made her "feel happier" along with the bond of sharing experiences with people who had come together because they were affected by the disaster.

People interviewed described food, clothing, and transport as being pooled among local people in most cases, although rations also were important and distributed on a communal basis. However, many people said the thing that people needed but could not get on their own was employment and vocational skills:

> We all lost our earning sources. We needed money to fulfill our needs, such as foods, houses, and all basic needs. If I am a fisherman, I have skills to catch fish. But if you ask me to dig soil as farmer, I can't do that. Maybe all jobs need different skills and it is hard to exchange. (Edy, street children worker)
>
> There was an NGO, that gave economic contributions for women so they could open a small shop, for example. (Novi, NGO worker)
>
> If you give us material goods, if we use it, it will finish. But not with skills. So, with skills, teenagers were able to look for a living. (Daudy, student and activist)

Several people said that aid distribution was uneven, with the rural areas and conflict-affected areas being less likely to experience assistance. Asrida, volunteer, describes planning as being a problem with aid:

> There is some overlap in aid...because there was no coordination between NGOS, not literally no communication, but a lack, so the aid is given again to that region and maybe there are some areas that haven't got any aid at all. That's from what I heard but the data, the real data, is unknown.

People mentioned that some people collected and sold off aid relief goods and used cash payments to buy gold (which people in Aceh often amass as a form of savings). Some areas such as Pucuk, being so remote as to need helicopter access and also being affected by conflict, were far more difficult to visit than Banda Aceh, where the aid was concentrated. Some local NGOs went to conflict areas for the first time after the tsunami.

Edy, the street children worker, described the insecurity of conflict and tsunami affected areas:

> I went to Idi, there were my cousins, nieces, nephews who had severe injuries. Broken bones, bruised, got flooded. Idi was still in conflict. The peace contract hadn't been signed yet. I heard the sound of gun shot several times. I had tsunami, and then conflict areas. I felt so insecure. I couldn't stay there, I wanted to go to Banda Aceh.

While there were many people who acted together after the tsunami, others said assistance made people lazy or devoid of motivation to do such activities as could improve their situation or that of their community. Fatma, NGO manager, said there were different responses to assistance:

> Some of that assistance I really needed. Clothes to wear, food to eat. But some people took that condition for granted. It means...there was this phenomenon where..."I got disaster, just continue helping me. I don't need any job, I get *jadup* every month [*jatah hidup* or living supplies]," some thought that way. So, it varied, some wanted that condition to end as soon as possible, some just wanted to take advantage.

Ahyar, community worker, pointed to the difference between smaller places like Calang, where help was not quick and people acted to rebuild, and the big cities, where many people would wait for help. This had an impact on work such as farming as well. Ahyar said,

> Most of them didn't want to work. Why? Because they knew that NGO will aid them. Lots of help, no need to work. They didn't even want to plant the rice. Take a look at Indrapuri, the area that wasn't struck by the tsunami, they didn't want to plant anything, because it was given freely. Mostly the people from outside were the ones who were doing all the work—from Java or Medan. One village could get help from 3 NGOs and they were smart, like: "Oh we haven't got any goods yet..." when in fact, they were hiding it and so they got it again.

People emphasized that those who were active rather than waiting for assistance recovered from the disaster better, and self-reliance and cooperation was seen by some people as important, as Dian, NGO coordinator, mentioned:

> I like how they handle the disaster in Jogja. The Sultan forbids the NGOs to come there. The Sultan asked people to work together to rebuild what's broken. So, the mutual assistance is still there. I wish they did it in Aceh.

Economic effects of the tsunami were considerable in Aceh and Medan, which both experienced economic growth with the influx of money and people into the area. Some development activities were poorly planned and had a go-stop effect. Asrida, volunteer, told a story of a local area given a salt house, but the price of salt was too low and so it was not used. In other instances skills training such as sewing was not supported by organizational and marketing activities so did not lead to increased earnings.

Economic benefits of tsunami assistance tended to favor those who were already well-off before the disaster. Erni, midwife, provided an example:

> One man, an NGO asked for his house: "How much?" "200 million rupiah a year." He rented it out, and then found another house that is smaller. Just like that until they are getting richer. But for the poor, they don't get any help, they don't have money to buy another set of clothes.

Aid organizations did create employment and high demand for local skilled staff with languages abilities, but many of those jobs were temporary—to cover the relief or recovery period and, when that ended, so too did many jobs.

Fatma, NGO manager, described how the temporary effects of aid organizations ended up having long-term cumulative consequences for the poor:

> We who'd never worked before then started to work: we who didn't understand this condition at all then entered this new world. I mean after the tsunami, the NGOs gained popularity. NGO, INGO [international NGO], everything. So, the people who were unemployed, college students, could all get a job, in INGO. Having an ability to speak English was important, but it wasn't the only priority. At the beginning, it was an emergency, everyone could've

been recruited, even if they were from the neverland. So, everyone did have the same opportunity, but we didn't know what was going to happen after. Then a problem began to arise when they started coming to barracks, giving this *jaduk* [supplies/subsidies]. We had problems with the place, with the helps given, with the family identity card, with the domestic violence, the inconvenience of living in barrack. So, the problems were all in one pack. Like the contributions, one person got it and the next one didn't: "I'm a tsunami victim, he's a tsunami victim, we're both victims, he's special and I'm not." So, there was some injustice at that time. So, generally it's like that in every place. But also prices jumped up, houses became expensive, and stuff. Gosh, Aceh was in disaster, and we can say that the ones who became victims were those who lived on the seaside, the coast, but those who lived close to the mountains were safe. But sometimes—it's not just my negative thought, it's reality—"I wasn't struck by tsunami, but I have to get advantages like the victims do," so, they got nice houses, they rent the house with high price. Times like that only last for a short time, emergency time. They were going to stay in Aceh for a few years then go, and it ends like this—Aceh eats its own relatives. After these INGOs were gone, the ones feeling the effects are the poor people.

Rumiati, local housewife, describes greed and those getting assistance even when not affected by the tsunami as negatively impacting people's relations with each other:

They didn't experience anything, they knew nothing about the tsunami, they hadn't seen even the water. But sometimes they took the biggest amount of stuff, and they didn't want anything to be given out if they wouldn't get it. That ruined our kinship. People discriminated against each other: "Oh you don't live here, you're just refugees, this is our place," things like that.

For Tabrani, NGO director, these problems were linked to the rise of consumerism and decrease in morality:

I'm saying before the tsunami, they had one house. But after the tsunami, they could get more than two houses. That's the irony. That's why I said the morality. The morality of the survivors. Another morality that I think was very bad is the rich people who rent the house for a very high price. It's really bad. I think because they knew that there are lots of money coming and they use this opportunity for getting benefit. Well, for me, hmm I say that for the local people that's very bad. Why? Because people from many countries came here to help us. That's bloody people. So, culturally because of a lot of donations, especially those for survivors, the attitudes, they become lazy and dependent with the donations. They use the money for consumerism. Actually, why didn't they use this money for building their life by running a small business? The lifestyle now after tsunami, Banda Aceh has been changed.

Opportunism also affected organizations as well as individuals, with organizations tearing down offices so that they could get money for rebuilding. Meanwhile, people were frustrated that despite the influx of assistance, some things did not improve, such as health services. Edy, the street children worker, complained,

> There are a lot of public health centers built by foreigners or by the domestic government. Such as Zainal Abidin general hospital, Meuraxa hospital, and the others. The facilities are supported by great buildings but not with staff services. I don't know why. The services never change, inefficient, and I'm disappointed. It happens not just at one public health centre, but almost all of them.

Sustainability of many aid programs and other interventions was another frustration. Asrida, volunteer, mentioned houses constructed by foreigners that no one knew how to fix (e.g., septic systems if they broke). Nurjanah, psychologist, said there were lots of job terminations as NGOs withdrew after tsunami assistance. Some NGOs prepared local workers for their termination at the end of programs, and some did not.

Relocation, living in camps, and building after the tsunami changed communities in both physical and social ways:

> In the past, there were not many houses at my village, so we knew each other better. But there are a lot of houses now, sometimes we don't even know people next door. Newcomers come to village. That's not a problem but I think it's better if we know each other. We are neighbors. It's good we have improvement on buildings development but it's more important to stay close with others. We don't socialize and communicate as well as in the past. There's no tight bond like before. I see it's disappeared. Before, we initiated community self-help and organized social activities. People could easily be asked to help. Now, it's so hard, it has already changed. (Edy, street children worker)

> In my village, it used to feel like a real village, the houses were few, but now it's kind of crowded, the traffic too, everything is getting more expensive now…and it is safer now…and the women are braver…lots of divorce now, in the religion's sector, in government's office, they deal

with up to few hundreds of them every month. That was something that could not happen in the old days. (Ahyar, community worker)

There are a lot of people who, at the first 2 months married, the tsunami widow and widower. On the 2nd and the 3rd month, maybe because it's a sudden decision, it makes a sudden separation. (Erni, midwife)

Most people agreed that women were worst affected because of the social expectations that they would care for their families and the loss of children, or manage difficulties of caring for children at this time. Violence against women was also a problem in insecure accommodation for refugees. Positive changes by aid organizations people mentioned included encouraging women to come to meetings and supplying their needs, including sanitary pads, medication for pregnant and delivering women, and general health services.

Nurjanah, psychologist, describes the burden on women at that time:

It's true that men work, but in that terrible condition, I could see husbands sitting on food stalls, daydreaming. But women couldn't do that, they still had to feed their children. She was then the one who thought about what they could do, like a woman at the barrack I saw, she washed clothes at different places. "Why is she never here?" "She's washing clothes," she made cakes, distributed them. She worked hard to feed her child, while her husband did nothing. Her husband could be disturbed for as long as he wanted, but she couldn't. "I'm sick, but I can't be sick for long. If I'm sick, what would my child eat?" even though if she had time, and we accompanied her, we consulted and gave her therapy, she would burst out. She was actually really fragile. She just tried to hide it all. It looked as if she was okay, while she was destroyed inside.

Erni, midwife, said children were also severely affected by living in camps without access to schools and places to play. The government allowing children to attend schools in places where they were (rather than where they were enrolled before the tsunami) helped some go back to school. Disruption of studies and education was a common concern for parents in Aceh following the tsunami. Daudy, student and activist, said one benefit from the tsunami was that more funds were put into education—previously, educational institutions relied on the government for funding, but with an increase in the number of funders and the amount of funds, activities on campuses increased dramatically, leading to better opportunities.

Apart from these economic and social effects, the tsunami assistance brought unprecedented interaction between people from Aceh and those from other parts of Indonesia and the world. This created several impacts, positive and negative.

People mentioned that while they appreciated outsiders' sympathy and desire to assist, the survivors of the tsunami sometimes said that it was difficult to determine what the outsiders were doing and why. One volunteer said mistrust was high and limited people's participation in aid activities,

> Like USAID, who made the watsan (water sanitation) program, actually, this watsan program should be done together with the people, so, the people also dig holes for pumps, so, the people will be paid too, meter by meter, but there are many people who don't want to do it because from stories I heard before there are some irresponsible sides—from the government which didn't give aid, people know that much of the aid is being affected by corruption so why work? Others work but do not receive aid because the fund has been corrupted. So, because they also know that there are many funds that should be for them but are not used for the sake of people's welfare so maybe they become dependent.

The lack of motivation in some people to help rebuild was attributed to NGOs paying money for people to do activities, so-called paid volunteers, and other individual payments that made people more self-interested than community-minded. Intan Dewi, program coordinator, Psychosocial Support Program, Aceh Province Office of Indonesian Red Cross, echoed this complaint:

> Now, everything was valued by money. Before, we helped everyone. As a Muslim, if we met another, we would be sure to help them, but now I saw this sentiment has started to fade. If another needed help, it was sure they would ask the question first: "You have the money or not?" I think it was something bad, that everything was valued by money.

Given that one of the key elements of Aceh's identity is religion and local culture, outsiders who did not understand this identity often created further mistrust, embarrassment, confusion, and sometimes conflict. Dian, NGO coordinator, explained that only some assistance was appropriate:

> There are aids that suit the people and there are aids that don't. Ya, the first one that suit the condition is giving people skill development.

They gave it to the exact target. But at that time, people needed to do some coping—they needed to tell their stories and share the problems they had. That assistance suited the culture here. But there was also a thing that didn't suit the culture. One time I saw, mothers were asked to dance and sing. It made them uncomfortable. But we just smiled that time. I think that those people just don't know that things like that are taboo here in Aceh. There was a dancing singing event in the *mushola*. Doesn't suit. There was also an issue, Christianization. Maybe it's just the perception, I don't know: Like books that use the letter Comic Sans, where the v looks like the cross: "Waaa, this is Christian." Things like that. And there's this one funny thing. They taught the children but they're Christian. But they force themselves to "Come on kids, read *Bismillahirrahmanirrahim*." When people knew that they're Christian, people thought there's political will behind it. Actually from what I see, Acehnese are not resistant to the aid from the non-Muslims. But just be honest, don't pretend to be a Muslim. I watched the local people and them fighting at Kampung Mulia. Local people said: "You don't have to lie that you're Christian," "Don't need to pretend to be a Muslim. You'll teach them wrong things," "Or you can hire Muslim to run your programs."

Nurjanah described how some assistance was not used because the local culture was not taken into account:

I was once given 1,500 dolls, and after sorting them, we could only use about 800. The ones that may not be used are dogs, pigs, and Santa Claus. They're really sensitive, so we got rid of them. Someone who brought the dolls to me asked: "Then where should I take these, ma'am?" "Don't return things and don't throw them away, because its people's gift, we have to accept it, thanks be to Allah. Just go to the Chinese barrack, just give it all to them." "Really?" "Yes." They must be happy there in Chinese barrack....We had an activity at Tungkup mosque, at the front of the mosque. We were playing story telling with the kids, using dolls. Then a Tengku came with anger, "...it's a statue, bla bla bla." He was saying it's an idolization. We started to get scared and backed down: "It's okay, relax. We keep the dolls here, but we can use it in other places." "Just don't play with dolls at the mosque."

Daudy, student and activist, mentioned that the thinking of some people was also rooted in history and experience of outsiders in the past, with people saying aid was a "second colonization." Language barriers also contributed to these problems of cultural competency and appropriateness, with Fatma, NGO manager, pointing out that even Bahasa (the national language) spoken by people from other parts of Indonesia could be problematic for people from Aceh.

At the same time, many people said that the tsunami assistance led to new relationships with outsiders and a certain "opening up" of Aceh to the outside world. This assistance coincided with a renewed effort to cease the armed conflict, which in many areas led to a feeling of greater ease than before. Intan Dewi, psychosocial support coordinator, said,

> Previously a lot of people have an opinion about Aceh people, they say Aceh people are one of resistance or persistence. But the reality is not like that. I see a lot of people now more flexible than before. Previously the reality was some of Aceh people were opportunist. Many of Aceh people have wrong paradigm, that every foreign people who work in Aceh is not part of them. They cannot adapt. But, later, it was changed, they started thinking that other foreign people also want to have a good life.... Many Aceh people have got a lot of psychological knowledge too. So, a lot of stigma about mental health was reduced. A lot of parents have started to understand about children's development.... The openness to people who weren't from Aceh, now we can see this not only in the town but also in the whole region. Before, Aceh people put on airs thinking that everything must be from Aceh people. Now, we have learned that we can learn from someone else. Now, we have evaluated ourselves and want to change.

Communications and transport infrastructure such as roads, radio services, ports, and the airport improved after the tsunami and also assisted Aceh's contact with the outside world.

People remembered the tsunami in a number of ways including prayer, ceremonies for the dead, sculpture, poles to mark the height of the tsunami, a museum, and dancing. Some of these remembrances were seen as inappropriate for local culture or premature because of the importance of Islam to the Aceh community, including the need to avoid idolization. *Kenduri* (communal feasts) and other religious services were significant ways of people getting together and remembering what had happened:

> The second month after the tsunami I was in Aceh Besar. High dependency, they waited for aid. Only when it was time for *kenduri maulid* [a ritual meal] did they act. This they did on their own, without any assistance from NGOs they made it. (Ahyar, community worker)

Many people said that the experience of the tsunami had led to learning, the most common comment being that the passing down of stories about the tsunami from generation to generation. would ensure greater awareness of disasters, as Fatma, NGO manager, detailed:

> There was a woman that was pregnant when the tsunami happened. She gave birth afterwards. Or another woman even gave birth in the water. It became like, my children will inherit this story, and they have to know, so this story will be told, and will run in the family as a history.

For others, the tsunami confirmed that Aceh people were tough and survivors. Edy, street children worker, described this sentiment:

> I think you already know about Acehnese. They used to face conflicts. They have a strong mentality to handle challenging situations. We have those capacities. We used to find ways to survive. Anything. We do everything to survive. All jobs.

This description of Aceh people as tough or "hard headed" meant that they also as steadfastly held on to traditions, despite the fact that some lifestyles had changed (mentioned as examples were unplanned pregnancies, liberal dress, and women going to coffee shops). Aceh culture in general, with its high respect for religion, and the street culture of coffee shops were seen to be the same, despite the tsunami and rebuilding. Dian, NGO coordinator, joked,

> What hasn't changed is Acehnese attitude who always go to the coffee shop, chatting all night long, building their castle there. The castle above their mouth.

CONFLICT, PEACE, COMMUNITY, AND DISASTER

While the peace process had begun long before the tsunami, the common perception in Aceh was that the tsunami made both parties more willing to make peace and intensify the peace process. So, many Acehnese say the peace enjoyed after the tsunami is linked to the common experience of the tsunami, drawing people together.

Regardless of the cause, the benefits of this renewed peace was felt by most people, and there were many comments of appreciation for this change. Ahyar, a community worker, explained the biggest change since the tsunami in this way:

> First, there is peace. You can feel that the most. Second, Aceh is free, so everyone can come now, every ethnic group. In those times, back

then, GAM hates the name that contains "O" very much even if it belonged to someone from Aceh. You could get in trouble because of that. The society is opening up because now they have seen *bule* [foreigner/white person]. And actually, with the abundance of aid, we could do a lot better than this.

However, certain shifts such as an increase in competition for money, materialism, and mistrust between certain groups gave some interviewees the perception that communities were less at peace among themselves than they seemed. Many people worried about the impact of loss of jobs and income as NGOs left Aceh and the memory of the tsunami faded in the international community.

The power shift from the people to government and aid agencies was seen as something that could influence peace, as Ida outlined:

> Aceh is more quiet, such as there were celebrations like the last Aceh Culture Week to show the world that Aceh has improved....There was so much aid given and it supported the society but not all the aid accommodated the society's needs, and not all people benefited from the aid also. For some people it has negative effects. Then, there are funds from the government...one day, without all this what about the Aceh people? So, because in general, they are still dependent, ya? Still not strong in the economy, every person, every community in the society I think that it's very important to make the peaceful condition, ya?

There was also anxiety about what had happened and about whether a future disaster would strike. Erni, a midwife, commented that it was easier to go out at night after the Memorandum of Understanding between the parties to the conflict, but the trauma from the tsunami remains:

> For me personally, when there's a heavy wind, whooo...there's a..."What is it?" "What's going to happen now?" like that. So if you said I have recovered from my tsunami trauma, not yet....But to me, I'm more steady in a way, nowadays, when we're doing our activities, there's no worries. There's no worries that GAM will shoot us. It's gone. Back then at Lampulo, after 6 P.M., I had already locked the house. We were alert at night. Right now, if somebody rings the bell at night, I think: "It must be someone who wants to deliver a baby."

Novi, NGO worker, said that time had made the tsunami survivors accept what had happened, and this contributed to a calm and a kind of preparedness to help deal with crises,

> After the tsunami I could work, I could help people, so I knew more because of what happened. Even if there's anything like this anywhere I really feel like helping. So that others wouldn't experience what I did. With no clothes, nothing. If we can help, why don't we?

Box 2.1   Recommendations for Interventions

Community members in Aceh had many comments and criticisms regarding the longer-term effects of aid. They pointed out that the effects of aid provided after the tsunami were often unintended and long lasting. Community members had specific recommendations to make for future interventions in Aceh dealing with natural disasters:

Let Locals Collect Data and Do Needs Assessment

- Ask local leaders for data on people and their needs.
- Coordinate data assessment within and between organizations to minimize redundancy in data collection (people being asked again and again).
- Involve communities to assess their own needs and make their own decisions about what needs to be done, with support from local organizations.

Acknowledge and Provide for Needs of Women

- Make camps safe for women by providing privacy and women's areas where women will feel comfortable.
- Make special arrangements of food and nonfood items (including specific needs such as sanitary products and health supplies) distributed in camps for women.

Make Aid and Presence of Intervenors Locally Appropriate

- Military personnel should respect and assist local people in this time of distress and need.
- Agencies should coordinate aid so that all areas where there are needs receive aid, especially remote and conflict-affected areas.
- Monetization of community activities should be discouraged (such as paid volunteers and paid activity participants) so as to not create a handout mentality.

- Interventions should be culturally appropriate in Muslim settings such as not requiring women to dance and sing (exception for traditional ones), providing toys considered idolization, or people other than religious leaders providing instructions for prayer and other religious activities.
- Services and supplies should be provided to groups rather than individuals, specifically food distribution and psychosocial support (rather than individual counseling). Interventions can then build on community cooperation rather than putting people in a state of competition with each other for supplies and assistance.

Take into Account Long-Term Effects

- Acknowledge and minimize longer-term economic impacts such as overly high new job salaries and price hikes that can create dependency and weaken local economies.
- Support recovery of livelihoods and skills training aimed at livelihoods in the postemergency phase so that people can rebuild their own ways of living.
- Provide skills training relevant to community-driven needs and post-training–related capacities (e.g., marketing, network expansion).
- Provide awareness and capacity building for local authorities to manage and sustain the assistance initiatives.
- Services should be funded—not just infrastructure—to avoid the situation where buildings are prioritized over staff, equipment, and other service needs.
- Interventions should be devised with an eye to sustainability rather than delivering services and supports that will end after one or two years.
- Measures that allow students to continue their education should be encouraged (such as the government's decision to allow students to attend schools in post-tsunami locations rather than going back to pre-tsunami locations).

REFERENCES

Brusset, E., Bhatt, M., Bjornestad, K., Cosgrave, J., Davies, A., Deshmukh, Y., Haleem, J., Hidalgo, S., Immajati, Y., Jayasundere, R., Mattsson, A., Muhaimin, N., Polastro, R. and T. Wu (2008), A Ripple in Development? Long Term Perspectives on the Response to the Indian Ocean Tsunami 2004, *A Joint Follow-Up Evaluation of the Links between Relief, Rehabilitation and Development (LRRD)*, Sweden: SIDA.

Carballo, M. and Heal, B. (July 2005), "The Public Health Response to the Tsunami," *Forced Migration Review*: 12–14.

Fritz Institute (2005), Recipient Perceptions of Aid Effectiveness: Rescue, Relief and Rehabilitation in Tsunami Affected Indonesia, India and Sri Lanka. Online. Available at http://www.fritzinstitute.org /researchCenter.htm, accessed April 21, 2011.

Grayman, J. H., Delvecchio, Mary-Jo, and Good, Byron J. (2009), "Conflict Nightmares and Trauma in Aceh," *Culture, Medicine and Psychiatry*, 33: 290–312.

Hedman, Eva-Lotta E. (2009), "Deconstructing Reconstruction in Post-tsunami Aceh: Governmentality, Displacement and Politics," *Oxford Development Studies*, 37, 1: 63–76.

Jayasuriya, S. and McCawley, P. (2008), *Reconstruction after a Major Disaster: Lessons from the Post-Tsunami Experience in Indonesia, Sri Lanka, and Thailand*. ADBI Working Paper 125, Tokyo: Asian Development Bank Institute.

Leitmann, J. (2007), "Cities and Calamities: Learning from Post-Disaster Response in Indonesia," *Journal of Urban Health: Bulletin of the New York Academy of Medicine*, 84, 1: 144–153.

Nazara, S. and Resosudarmo, B. P. (June 2007), *Aceh-Nias Reconstruction and Rehabilitation: Progress and Challenges at the End of 2006*, Tokyo: Asian Development Bank Institute.

Ride, A. (1999), "Indonesia," *New Internationalist Magazine*, p. 318.

Telford, J., Cosgrave, J., and Houghton, R. (2006), *Joint Evaluation of the International Response to the Indian Ocean Tsunami: Synthesis Report*, London: Tsunami Evaluation Coalition.

United Nations Development Programme (2006, 2007, 2008, 2009 editions), *Human Development Report*, New York: UNDP.

Wandita, G. (November 2008), "The Tears Have Not Stopped, the Violence Has Not Ended: Political Upheaval, Ethnicity, and Violence against Women in Indonesia," *Gender and Development*, 6, 3: 34–41.

World Bank Office, Jakarta (January 2008), *Aceh Poverty Assessment 2008: The Impact of the Conflict, the Tsunami and Reconstruction on Poverty in Aceh, Indonesia*.

3

PAKISTAN

Asha Bedar, Diane Bretherton, and Anouk Ride

Dr. *Asha Bedar is a clinical psychologist, group facilitator, trainer and researcher, and author of various community-based modules and manuals on issues related to children and women's psychosocial and emotional health, violence, and gender. She is currently working with an NGO based in Pakistan as well as a community welfare organization in Australia. Originally from Islamabad, Pakistan, she now travels and works between Islamabad and Karachi, Pakistan, and Melbourne, Australia.*

Ali Asghar Khan, the executive director from the Omar Asghar Khan Development Foundation (OAKDF), an NGO dealing with development, democracy, and humanitarian assistance, described the shaking of the earthquake in his house with a laugh: "It was like the Exorcist! My bed was like the Exorcist! The way it was shaking."

Rashida Dohad, his friend and programme director from the OAKDF, agreed: "I could not open the door because it was rattling so much. You couldn't hold the handle, it was like that."

And that was in Islamabad, a distance from the epicenter of the earthquake, Rashida pointed out:

The stories one has heard from people who were closer to the epicentre—they said that you know like good, strong trees were swaying, I even can't imagine what it was like closer, and we've also heard people very close to the epicentre saying that there was so much dust that initially they couldn't see what had happened because of the collapsing buildings and landslides, because of the mountains and all.

It is the terror you feel because you can't even depend on the solid ground you stand on because the ground shakes so much. So, it's really terrifying. You know I've worked in Hazara for years and years, but I think for a long time I was afraid of the mountain roads, which I had never ever been afraid of. I have worked there, I have walked in the places, I have driven in very difficult circumstances, difficult roads and everything else, but I think for months that fear stayed with me of the roads.

Ali Asghar: You were getting, you know, aftershocks about 15, 20 times a day you know and some of them were quite strong...

Rashida: Quite strong.

Ali Asghar: I mean, my family was sleeping in the car. They didn't want to sleep inside.

Rashida: You know, as we did all the relief work and what we did was, which was not really planned as a psychosocial support or anything, but we decided to all stay in the office. So we used to live there, that was where the warehouse was and since it was Ramzan we decided to give meals to the staff because they were all going for, you know, nine or ten hours. When they would come back there would be very forced laughter. We had all heard, terrible, terrible stories of absolutely devastating things that had happened to people. And then we realized and sort of, noticing differences in behavior and the way they were relating. Those shared meals I think was a form of proving support.

What was absolutely fabulous is how ordinary Pakistanis came forward, the rich, the poor, middle class, and a lot is said about that, which is very well-deserved and I found it extremely humbling the way people responded. But, I think, still the stories that are still not well-known are of the survivors themselves and how they too played such an incredible role in helping each other, whether it was sharing goods or sharing food or just the solidarity of grief, so you know together dig graves and stuff like that.

Ali Asghar: You know there were heroes at every point...

Rashida: Every point.

Ali Asghar: ...from the survivors, and they were organizers, you know who were organizing things from the beginning.

Asha Bedar (Interviewer): So the initiative was coming from within the survivors.

Rashida: Yes. This is less known, less known.

Ali Asghar: That was sometimes tapped...and mostly it wasn't.

Asha Bedar: Because we would hear a lot about the response of the civil society and aid and volunteers, but their own effort isn't documented?

Rashida: And, you know because most were from rural communities I think many did not know the extent of devastation, so when you had people coming from Karachi or something and you know somebody would say...and they would feel so grateful that "my

God, they've come from Karachi to help us!" And the basic generosity and hospitality, so you would find that even a person who
has lost everything but still able to put some *pakoras* [a snack: gram
flour fritters] together because it's Ramzan and all for the guests
who have been so kind to come all the way from Karachi. That, I
think, was partly because of the isolation that they live in. So how
big a deal it was and the fact that they were on CNN and BBC was
completely lost to them.

CONTEXT

Before recent disasters and the "war on terror" came to Pakistan,
it was known for its extraordinary founding, being established
as a Muslim nation in the wake of the dissolution of the British
Empire. Officially the Islamic Republic of Pakistan since its first
constitution in1956, it is situated in South Asia and shares borders
with Afghanistan, China, India, and Iran. The country comprises
four provinces (Sindh in the south, North-West Frontier Province
[NWFP] in the northwest, Punjab in the east, and Balochistan
in the west), the Federal Capital Territory (Islamabad), and the
Federally Administered Territories and Areas (FATA). Pakistan also
has jurisdiction over the Kashmir region, known as Azad Jammu
and Kashmir (AJK) in the north.

Across Pakistan, relatively high population growth has contributed
to loss of forest over time as well as increased landslides and flooding.
The north of Pakistan is subject to earthquakes and seismic activity
due to its location on the Iranian plateau at the junction between
the Indian subcontinent and Eurasian plate. Here, the snowcapped
mountains contain glaciers and are colloquially known as the "third
pole" with the world's second-highest mountain (K2 at 8,619 meters).
However, in summer, the region also experiences extreme heat and
sometimes drought-like conditions.

It was this region that was hit by the earthquake of 2005. The
most severely affected areas were parts of the NWFP (the Hazara
region) and the AJK.

While Pakistan's economy is dominated by services, accounting
for around 50 percent of the GDP, farming the land still accounts
for 20 percent. In the rural areas of Pakistan's north, most people are
farmers and foresters with cash crops including wheat, maize, rice,
sugar beets, and some fruits such as apricots. Here, people are considered rich if they own land, brick housing, cars, gold, and livestock.
But these assets are also tenuous, as many people have experienced

extreme climactic conditions, earthquakes, droughts, floods, and landslides affecting their agricultural income.

However, the rural-urban divide and glaring socioeconomic inequalities make any generalizations about the economic conditions of the people in the north near impossible. Much of the rural population has little education and lives in the mountains in small mud houses with large families living under one roof. Yet, a small part of the population in both urban and rural areas consists of business people, landowners, and educated professionals, living in large concrete houses, with high income levels and domestic servants.

Overall, however, Pakistan is a developing country that faces problems with high levels of poverty and illiteracy. It is also the sixth most populous country in the world and has the second-largest Muslim population in the world after Indonesia. Pakistan ranks 136th on the United Nations Development Programme's (UNDP) Human Development Index, below India but above Bangladesh. It has a life expectancy of 64, around ten years less than China's life expectancy, and a higher mortality rate for children under five than India (99 per 1,000 births, UNDP, 2007: 255). Public health spending is just 0.4 percent of the GDP (UNDP, 2007: 249).

Prior to the 2005 earthquake, Pakistan had around $14 per capita aid a year (UNDP, 2002: 200). Political issues such as the military regime's nuclear program and the conflict in Afghanistan restricted aid flows to Pakistan up until the United States' "war on terror" when the United States came to view the government as an important ally and changed its aid policy. The United States began to spend around $600 million each year on aid in the country, with others following suit, and boosting aid to Pakistan to over $2 billion in 2002 (this dropped back to $1.6 billion by 2007).

Pakistan's status as a Western ally sits oddly with its combination of a British parliamentary system and military dominance of politics and the bureaucracy. The government of Pakistan has an indirectly elected president as the head of state and commander in chief of the armed forces and an indirectly elected prime minister as the head of government. Governance in Pakistan has been tumultuous, with military regimes interspersed with elected governments since independence in 1947. Numerous dissolutions of elected legislatures (from 1988 to 1999 the National Assembly was dissolved three times by the president and then by a coup d'etat) have created a certain lack of trust in politicians and government by those not part of the network of privilege.

In addition to this, there is conflict due to the politicization of religion. Around 75 percent of the population is Sunni and much of the remaining percentage comprises Shia Muslims. Religious minorities include Hindus, Christians, Sikhs, Ahmedis, and Zoroastrians. In the last two decades, Pakistan has seen a rise in intersectarian conflicts and militant "jihadi" groups, which warred among themselves, with 345 people killed in sectarian fighting from January 1997 to October 1999 (Haleem, 2003: 469).

Pakistani writer Ibrahim Malik argues that despite economic development and urbanization, "Pakistan is still a predominantly agrarian, rural and feudalist society." Apart from land, previously prestige was also gained through acting as intermediaries and establishing political and economic relationships with British colonial powers. However, today, landed elites, and the middle class, build their power bases through positions in the bureaucracy and military, or through marriage, doing business, and creating political allegiances (Malik, 1996: 679; 1997: 92–93).

The north, most affected by the 2005 earthquake, also experiences these ethnic, religious, and power divides—with the majority of people, Sunni Muslims, speaking the Pashto language. There have been some moves to rename the simply titled North West Frontier Province (NWFP), which runs 1,100 kilometers along the Afghanistan border, to *Pakhtunkhwa*, "Land of the Pakhtuns." The region has popular histories of poets and warriors often reinforcing ethnic identities and militantism. This is further complicated by the British division of Pakhtun tribal areas along the Afghan border, the ongoing conflict between India and Pakistan regarding the sovereignty of Kashmir, plus the previous influence of the Soviet-Afghanistan and the current Afghanistan "war on terror" conflict (the latter two led to an estimated 1.5–3 million Afghani refugees making their home in the north).

These conflicts and the tribal tradition of "honor codes" (behavior limitations on women to preserve men's honor often enforced through violence) have led to the marginalization of women, particularly in the NWFP where women face social restrictions on movement, behavior, and relations with other tribes and clans. Both men and women uphold these honor traditions, including in some cases killing or disfiguring women who are considered to have brought shame on the family (Ahmed, 1997: 44). However, despite this, women are prominent in NGOs in Pakistan, and the late Benazir Bhutto, advocating reduction of the power of the military in government, was the first woman to be elected as a prime minister of a Muslim country.

Lack of governance and military power has fed each other as Irm Haleem (2003: 474) from Seton Hall University in New Jersey states: "Poor governance, economic disparities and fragmented society, eliciting alliances for personal not collective gains (either on the part of ethnic or sectarian parties, civilian or military elites), have increased the saliency and influence of the military in Pakistan."

There are an estimated 10,000 NGOs in Pakistan, many growing out of the Afghan war in the 1980s when donors shifted toward funding NGOs (Islam, 2001: 1343) in response to the "Islamisation" of Pakistan under the Zia regime and the resulting discrimination against women and minorities (Mumtaz and Shaheed, 1987). Since the Zia regime, a number of political/religious *jihadi* (religious war) groups have established themselves in Pakistan. It is estimated that there are about 58 religious political parties and 24 known militant groups. The religious political parties also have militant wings or maintain links with local and regional *jihadi* networks, most operating from the NWFP. Since Pakistan's alliance with the United States there has been a clamp down on terrorist groups, especially in 2009–10 as the conflict with the Taliban has intensified and Pakistan has faced a series of terrorist attacks on its soil in retaliation, with thousands of civilians killed.

Removed from the country's political and economic power base, in the mountains of the NWFP, people are viewed as more traditional, insular, and "rigid" or "stoic" in their views than in other more developed and cosmopolitan parts of Pakistan. In this region, people's anxieties include urbanization, family breakdown, increasing difficulty to make ends meet, and lifestyle changes, and these have been found to influence how communities perceive and respond to risks (Halvorson, 2003: 276).

According to the 1998 population census, there are 88 people in rural areas for every 12 people in towns and cities in AJK. The majority of the rural population depends on forestry, livestock, agriculture, and informal jobs. The literacy rate in AJK is 64 percent, one of the highest in Pakistan. Generally, the AJK is considered to be more open, with better human development indicators.

Since the earthquake, armed conflicts between the Pakistan army and the Taliban have once again pushed parts of the NWFP and FATA provinces in the north into the midst of a humanitarian crisis, with over 3 million people displaced within Pakistan in 2009, the highest number of Internally Displaced Persons (IDP) in the world at that time.

On Saturday October 8, 2005, at 8.52 A.M. Pakistan Standard Time, another crisis came to Pakistan—a major earthquake measuring a magnitude of 7.6 on the Richter scale hit parts of the NWFP (the

Hazara region) and the AJK (Pakistan-administered Kashmir), near the city of Muzaffarbad. It was also strongly felt in and affected parts of Pakistan's capital city of Islamabad. It is believed to have been one of the most destructive earthquakes of all time, resulting in an estimated 73,000 (official figure) to 100,000 people killed, over 4 million left homeless, and 8 million affected. The severity of the damage caused by the earthquake is attributed to severe upthrust, combined with poor construction. Entire towns and villages were completely wiped out in parts of Pakistan, with other surrounding areas also suffering severe damage. Many government buildings collapsed—over 6,000 schools were destroyed, with schoolchildren being one of the largest group of causalities (17,000 killed and 23,000 disabled). Eight-hundred clinics and hospitals were destroyed by the quake, as well as 400,000 houses and nearly 4,000 miles of roads, often blocked by landslides. Communication networks collapsed and agriculture was disrupted, with 250,000 farm animals perishing and, at the onset of winter, food storage and agricultural facilities were damaged leading to food shortages in the region, with 2.3 million people categorized as "food insecure" by the World Food Programme (UN, November 2005).

RESEARCH OUTLINE

To find out how people coped with the earthquake, we conducted a total of nine interviews with ten people (as Rashida Dohad and Ali Asghar Khan of the OAKDF were interviewed together). Five men and five women participated. Four of the nine interviews were conducted primarily in English, three were conducted in Urdu, and three were conducted in a combination of English and Urdu. The participants were recruited through purposive sampling (selecting people based on categories and criteria). An effort was made to interview not only a diverse group of people largely from the NGO/community/development sector, but also the business community, volunteers, the media, and human rights and religious groups. All the participants had either worked directly with the community in the affected areas (e.g., volunteers or community organization workers) or had worked with groups/organizations that were actively involved (coordinating or managing volunteer/psychosocial support/development efforts). The primary researcher, Asha Bedar, is also a part of the NGO sector and was actively involved in a psychosocial support project after the 2005 earthquake, so some of the participants were recruited through personal contacts and others through referrals. Decisions about whom to interview started by talking to a couple of fellow NGO colleagues and brainstorming with them around potential intervie-wees, keeping in mind the criteria, diversity in their fields of work and perspectives, active involvement in relief/support work, and reliability of information. Several of the participants identified other potential interviewees, some of whom were then contacted, and the process was repeated. Though a large number of potential interviewees were identified, ultimately, the list was limited to ten based on who was most reliable in terms of their involvement and overall perspective, as well as on who was most available in the coming months, especially because of their work in the ongoing IDP situation in the north of Pakistan.

COMMUNITY RESPONSES

Earthquakes are not uncommon in Pakistan, but the magnitude of the 2005 earthquake was so great that it was not something anyone had ever experienced or even heard of. Not surprisingly, the immedi-ate reaction reported by most people in both the most significantly affected areas was of disbelief and shock. Some people reported their minds going blank and freezing, initially not even thinking to run outside and save themselves.

In Islamabad, people were horrified as the Margalla Towers, a modern ten-story high residential complex of luxury apartments in a wealthy locality, shook and crumbled, killing more than 100 of its residents and seriously injuring others. To people in Islamabad this was a sign of the magnitude of the earthquake and was the first bit of earthquake news that dominated the media. As local journalist Abdur Rauf, who later worked in AJK, noted: "Nobody could have imagined that this would happen."

It was a chaotic day in Islamabad characterized by rescues, loss, injury, and a general sense of anxiety and sadness. Many reported rushing to and from hospitals, desperately calling up to check on rela-tives and friends living in or around the Margalla Towers, and a mix-ture of utter disbelief and helplessness at the emotional and physical proximity of the damage (and in many cases personal loss). It was not until late afternoon, when the news of the scope of the earthquake began trickling in, that Pakistanis became aware of the extent of the devastation the earthquake had caused in the rural areas of the NWFP and AJK. Many reported being "glued to" their TVs as they watched the crisis unfold before their eyes live on their screens. The images of critical injuries, dead bodies, people buried under the debris, flattened towns, collapsed schools, and so on, were overwhelming and strongly embedded in the minds of those who watched. As Maria Rashid of an Islamabad-based NGO, Rozan, said,

> I personally remember being severely depressed by it because there were just...by the evening we were inundated with images on TV, of children, of buildings collapsed, news of schools collapsing, lots...a huge number of children being affected. So just being very emotionally sort of...what's the word...upset, almost...no in tears literally. I remember I used to watch TV...for the next couple of weeks I would watch TV every night and invariably end up crying.

The AJK province and many parts of the NWFP province were severely damaged, and many villages were completely destroyed on that Saturday morning. Buildings crumbled, trees fell, thousands of people were buried under the rubble of their homes and schools, towns were flattened, and severe landsliding was triggered in the mountains. Saturday, being a regular working day in the public sector, meant that most children were already at school. It was also the Islamic month of Ramzan, so most people were fasting during the day and were taking a nap inside their houses after their predawn meal, leaving them little time to escape. The areas most affected by the earthquake were also some of the most economically and geographically disadvantaged, including certain remote mountain villages that did not even appear on local maps and to which there were no developed roads. It left everyone in awe:

> *Tahira Abdullah, Rights Activist*: Balakot...of course everyone knows the whole city of Balakot went down. All the way up to Kohistan, Balakot, Battagram, Alai, Shangal, all these places, severely affected. So, the estimated figure is close to about 150,000 people dead.
> *Abdur Rauf, Journalist*: A lot of students you know were buried in their schools. *It had completely collapsed.*

That the early hours and day were characterized by chaos, panic, and distress is understandable and undisputed. However, the dynamics of the community response—its extent, causes, and implications as well as whether, when and how the community responded constructively and collectively—have been interpreted differently by various community members, workers, and observers. Naila Azam, then a teacher in a moderately affected town of the NWFP, talks about the typical response to the terror:

> I was taking my Chemistry class and suddenly everything...the building started shaking and slowly the tremors began to get faster, and then all the children started running outside, I also ran out and then some children fainted....And then it took a while to take care of them. And the children could see that cracks had developed in the walls and the 4-storied building in front, that was our college section, was also swaying. Then all the children gathered in the ground and there was complete chaos...so then controlling them and watching over them so that they didn't get lost in this sort of a situation. It was a horrible experience...of 2005.

The screaming, shouting, crying, running, and praying was reflected in the media and in the stories of many of the survivors, workers, and observers. But in the midst of the chaos, there was also an immediate sense of shock, sadness, and helplessness in the aftermath as the utter enormity of the loss sank in for some people. Ali Asghar Khan of the OAKDF noted,

> If you look at the people who were in touch with, people who were actually, one can say the real victims of the earthquake, one found that they were...I mean...in a sort of a trance...inactive...you know?

Some of these people were unable to respond to the needs around them, so overwhelmed were they by their own distress, loss, and challenges. Ali Mujtaba, a volunteer in AJK, recounts a time when he saw funeral processions of people whose lives had been lost go through town without any of the onlookers stopping to lend a shoulder to the shrouded bodies. In a culture where attending and helping out at funerals is considered one of the most important social and religious duties, this apparent apathy shocked Ali. It was only later when he became involved in a psychosocial support program with a local NGO that Ali was able to see that these people were too traumatized to move.

Yet, at the same time, inactivity was not a typical response. Many people were seen to be highly proactive and immediately jumped into action. Numerous stories were told of local people involved in immediate rescue and relief activities, including those who had lost their own family and loved ones. As Rizwan Ashfaq, a worker with Islamic Relief Pakistan, in AJK reports,

> Well, as I told you there was shock and panic. But then I think people who hadn't had human losses immediately started helping

others and I know some people who had dead bodies of their family members under the rubble of collapsed houses, but they were still trying to save lives of other people who were still alive but trapped.

Community perceptions of the earthquake varied. An immediate thought that entered the minds of the religious-minded, panic-stricken people of the affected community was that *kayamat*, the Day of Judgment had arrived. A strong religious leaning in those regions, in particular, meant that the earthquake was immediately labeled by some as retribution for the sins and immoral behavior committed by the people of the region. Rumors began flying almost immediately of another more destructive, impending earthquake, causing more panic in some communities. But amid the distress, there was also a sense of calm and acceptance demonstrated by some people, accepting the earthquake as the will of God, thanking Him for what had *not* been lost. Ali Asghar Khan expressed a sense of surprise when he asked a man who had suffered substantial loss in the earthquake how things were and the man responded, *"Alhamdulillah"* (thanks be to God), meaning that all was well. Such comments reflect a deeply religious outlook on life according to which complaining about their losses, seen as the will of God, would be disapproved of. Instead, the man accepted his loss and thanked God for his blessings, an attitude that in the face of this crisis, helped many people move on with their lives.

The community's own efforts in terms of supporting themselves and those around them varied, it seems, from region to region and from one stage of the crisis to the other. Rashida Dohad and Ali Asghar Khan expressed a strong view that within the community "there were heroes," and that the stories of how the communities helped themselves were lost. Others such as Naila Azam and Rizwan Ashfaq echo this view, emphasizing the involvement of the community. The immediate response of communities everywhere was to make sure their families and friends were safe, to rescue their loved ones who were injured or trapped under the rubble, and to identify and bury the dead. Rizwan Ashfaq, from Islamic Relief Pakistan, observed in Muzaffarabad,

If they suspected that there was a child or a woman or a man buried under a house, who was still alive and there was a chance for rescue, all the neighbours would get together and start digging and do whatever they could. The people had no technical knowledge, no equipment, but they would do whatever they could together.

The people tried to do what they could to save lives as Tahira Abdullah, a human rights activist, described:

> People were using their hands. I've seen people using their bare hands…in the Margalla Towers and they couldn't do anything because there were huge chunks of concrete. Bare hands are useless.

As the programme director of OAKDF Rashida described, people looked out for each other:

> What we find very fascinating that sense of cohesion, that fabric was also in a sense…because people were so traumatised that they started thinking just for themselves. That rural cohesion in many cases frayed, in many cases stayed together. When it came, for example, to relief items, when people started giving things on a household basis, you would often find people taking it for their household, which was in a sense eroding the rural cohesion, but on the other hand, there were also many cases where you would find that they would share their grain. Because it was Ramzan, what happens mostly in rural communities is that they stop working for the month so they don't have to keep coming back and forth to the mountains because they're fasting, at the beginning of Ramzan before the fast starts it is traditional to stock up. So, if you had other families that have lost their stock or the stock was buried under debris or something, you would find that, you know, that there would be one sharing without considering how they were going to survive the remainder of the month.

This sense of thinking beyond one's own personal loss and needs and looking out for those around them was also described by Ali Mujtaba, volunteering in Kashmir, as he talked about a man whose house had been lost in the earthquake, but who took on the role of cooking and providing food for the affected community and volunteers in the initial weeks. Similarly, Naila, a community worker in the NWFP province, described how the earthquake demonstrated solidarity that was not always acknowledged once the work and media coverage of the international organizations started:

> Aid organisations didn't get there immediately. The people did a lot themselves too. Of course, when people's energies had been exhausted and they couldn't do much else, that's a different thing, but the solidarity that was there. There was no separation between my house and their house; it was a shared pain and grief. People from here, from Haripur, from the Punjab, from Karachi were getting relief goods there and the people there were distributing it. This is how it was. At

times, we did hear that somewhere vehicles carrying relief goods had been looted, but people from within that region didn't do that, this was done by other elements from other parts who were looting them on the way.

Many shared the view that this solidarity, sense of cohesion, and community efforts, which continued into the recovery phase, such as communities spending their own money to get things rebuilt, were rarely acknowledged, particularly when the media coverage of the international organizations took over. As Rashida of OAKDF observed,

> But I think still the stories that are still not well-known are of the survivors themselves and how they too played such an incredible role in helping each other, whether it was sharing goods or sharing food or just the solidarity of grief, so you know together dig graves and stuff like that.

Abdur Rauf, a journalist covering the crisis in AJK province, noted that this pattern of helping others and supporting relief efforts even in the poorest and most affected areas also extended toward the volunteers and organizations that came in to help:

> And naturally, I mean people in that area are quite poor, but surprisingly...I mean in Bagh, when we reached Bagh, there were only a few houses that were still intact...we took with us some tents, but people were so hospitable. I mean houses that were intact, they wanted us to stay in their houses, they brought us food, they brought us, I mean, you know, all the hotels and restaurants were closed, so we took canned food with us, but there was no bread. So, people made bread in their houses and brought us bread. So, even in that situation people were so hospitable.

At the same time, in some communities a certain shock or apathy set in. Ali Mujtaba, a volunteer and former boy scout, described in Azad Kashmir that

> there are some people who are just lazy, but there were some people who were completely numb.

It seems communities that were already organized and had previous support from local organizations acted quickly and more constructively, focusing on dealing with their needs immediately and that this, in turn, meant that in the future they were better-off psychologically than those who waited for assistance. The difference in self-reliance in the early days was also linked, at least in part, to regional and cultural differences in the two major areas affected. As Tahira Abdulla stated in describing the collective response of communities,

> Hazara and the North-West Pakistan are very, very conservative. Certain areas of Azad Kashmir are conservative, but most areas are not. They were very progressive, very liberal, the literacy, the education rate is much better, the health and education indicators are much better, the gender indicators are much better, so the...progress and development in Kashmir are much better than that in NWFP. Everyone knows that. As a result, the psychology and the ideologies [to formulate a collective response] are much better in the Kashmir region than they are in the NWFP.

The level of volunteerism—not only by people who were not badly affected, but also by affected people themselves—was also strongly highlighted by community members. Rashida Dohad of the OAKDF told just one of many stories of the affected people who helped others:

> There was one gentleman who was working with us as a volunteer. He was in our office one day and one evening we were having dinner and he mentioned, he just said, "You know, my son, I haven't found him yet." I was shocked! And the son was in Balakot, and had gone there for a job as a hotel waiter. Fortunately, after many weeks he was found alive and well. But for many weeks this man did not know whether the son was alive or not, and he was volunteering his time for his remaining family, you know the village. He himself was 75 years old, he was right across the epicentre, he had seen his own home fall. And there he was, he mentioned it in such a...almost a matter-of-fact way, that "you know my son I haven't found him, I don't know where he is."

The level of volunteerism from the larger Pakistani and international community also emerged as a strong achievement of the time. Maria Rashid, the codirector of Rozan, a local NGO working on emotional health, women's issues, and against violence, in Islamabad, describes the outpouring of sympathy to people affected by the earthquake at that time:

> I think everybody was, very, very upset, very affected, very upset. A lot of the next one or two days people wanted to do things, started volunteering, started raising money, informally.

For a volunteer from the Karachi business community, Agha Murtaza, it was a patriotic feeling that prompted him to use his contacts and money to organize and support efforts such as organizing the warehouse of supplies, finding and funding doctors to go to affected areas, and getting them to callback what the needs were for supplies and medicines, which he said filled up his house within a day. In Muzaffarabad, AJK province, he circulated in an old, borrowed car with a sign outside that said: "I have shrouds. If anyone needs any, wave your hand and I'll stop the car and give it to you," and later he was involved in building houses for widows.

In Islamabad, the Joint Action Committee (JAC) formed an emergency response committee (JAC-ER) to raise funds, recruit volunteers, and organize the efforts of different NGOs. Tahira Abdullah, one of its most active volunteers, describes the JAC-ER as handling volunteers, advisory services, food, medicines, health, education, donor relations, UN liaison, finance, and purchases. Tahira remembers being overwhelmed by volunteers wanting to help out:

> We set up a database of volunteers coming from all over Pakistan. By the way, while I had middle-aged and old-aged volunteers from Pakistan, the Pakistani youth is something so admirable...it was something that...I think the most hardboiled of cynics and sceptics about the future of Pakistan—and there are plenty of those around— probably had all their hardboiled crustiness shattered at the time of the earthquake in 2005. They came in their 10s of 1000's. For every foreigner we got we had 9 Pakistanis coming in.

Maria Rashid of Rozan supported this and described how they had to initially run daily orientations and training sessions for volunteers wanting to go up north to offer some level of emotional support. Achievements of volunteers and community organizations include donations, convoys of supplies, medical volunteers and volunteers at hospitals, training of volunteers in psychosocial skills for hospitals, and volunteer activities with children.

This public response also was not without some problems. Local journalist Abdur Rauf in Azad Kashmir described how roads were choked with people who had hired trucks to go north and deliver supplies, some took food that then perished on the journey, some took clothes unsuitable to the people and climate (so people hoarded donations to then sell to the market). Thus, while there was a strong collective desire to help, the understanding or the effort to understand what the requirements of the affected communities were at the time was sometimes missing, which created problems for relief efforts at times. Agha Murtaza, a volunteer from the business community, describes this:

> Once the aid started pouring in it was like a Pandora's box. Trucks of water, trucks of juices, trucks of things which we don't...they don't even need, for instance chocolates! People donated their clothes, and I was astonished and I took pictures of every single turn on the road that had piles of clothes just lying there and vehicles running over them. You can't even imagine—people donated their ties! You know the shirt tie? What the hell do you think a person living there would do with that tie?

While the efforts of the community and the outpouring of support from volunteers from around Pakistan were praised by everyone, many people felt that there was a sharp decline in these efforts with the arrival of aid assistance from outside the community. As the crisis approached what should have been the recovery stage, a certain level of dependency, competiveness, and dissatisfaction began to set in as the affected areas were flooded with relief supplies. Rizwan from Islamic Relief Pakistan described the change in AJK province:

> Of course, after a few days, when we go there, to be honest...when INGO [international NGO] sector people go there to distribute relief goods, after that, you could say that the first week goes very well in any area where there has been a disaster—everyone looks after each other, they say, I have this, so give it to so and so. But after a week, everywhere—the same thing happened in AJK or Mansehra, people discover that supplies are coming. Once they've been fed, their greed surfaces. After a week, it's like let me take this and let me take that. Then they start lying and all that. But this is a separate thing. In the first week after any disaster, anywhere you go, in the first 2 or 3 days people help each other a lot.

While everyone was affected in some way, women were seen as one of the most affected and vulnerable, with some people also mentioning the vulnerability of children. More women than men were reported to have been killed in the earthquake because of the gender dynamics and social conditioning in the region, as Tahira Abdullah, human rights activist explains:

> A lot of the women, where there were men, they ran out and saved themselves. A lot of the women ran back to fetch their elderly father-in-law, the elderly mother-in-law, people who have rheumatism and arthritis, who may not be able to walk, who are in a wheel chair or too

poor to own a wheelchair and are just in bed. Now isn't that a telling circumstance? And more women died. The government is trying to hide that.

Widows were also seen as a particularly vulnerable group because of the low rates of literacy and lack of vocational skills, resulting in a high number of dependent female-headed households. The sense of helplessness and dependency caused by gender dynamics was further reflected in the case of disabled women, many of whom were divorced or abandoned by their families and forced to rely on support from NGOs and volunteers. Their option of a good quality of life on their own was limited, as business volunteer Agha described when telling the story of a young woman he rescued from the rubble:

> Luckily I saw her. I thought it was a dead body, but then I saw some movement, I yelled. So, when she was rescued and everything, after a month I tried to call and ask a few people and I was told she was in a particular hospital, so, I went there, especially to meet her just to, you know, see, you know...you're the one I saved. When I met her and told her I was the one, she started crying and she started cursing me about why I had saved her that I should have let her die over there. I was disturbed, I felt bad. After she calmed down I asked her, "Could you please explain why you said that to me?" She told me: "You people from land areas don't understand the culture of the mountains. I'm not a beautiful girl, I'm a very ugly looking girl I know that. The only reason why anybody would marry me is that I would be able to come down from the mountain, collect wood and go back. To some extent, for sexual pleasure as well, and children or whatever, but those are secondary or even third-level issues. The first issue...the first only reason he would marry me is that I would be able to work. Now I don't have a leg, so I've missed that chance as well. I'm just 19 years old, nobody will marry me, nobody will take care of me and I will die like this."

In other cases, young women from Kashmir and Hazara (who are known for their beauty) went missing from hospitals. Often women, who could walk around the village, could not go outside the tents because of the *purdah* (veiled or segregated requirement), so they were confined, and there were reports of domestic violence in the tents.

When it came to assistance, many described how, because of the segregation, women were often reluctant or unable to come forward on their own. Development worker Samina describes how in the rural areas women were afraid that they would be considered "fast" if they approached strangers, so they hid in the cornfields and sent children to go get relief supplies for them. They often received less, as community worker Ali Mujtaba described:

> For any relief goods they had to queue up and in every queue there were men who wouldn't let them [women] move up the queue. Similarly, we saw that the more men a household had the more relief goods it was likely to receive. The more women a household had the more likely they were to be vulnerable. And see with children somebody or the other always gives them attention, but no one was ready to listen to the concerns of young girls.

Tahira Abdullah gave yet another example of the conservatism of the area and how women's issues needed special attention:

> Because of the conservatism the men used to take the pack... take the family packs, which had every thing, including sanitary napkins and they would take the sanitary napkins and throw them on the roadside and say, this is anti-Islam, Western... immoral stuff, *fahaashi* [vulgarity, immorality, obscenity].

People in rural areas were also at risk of poverty, as Rashida of OAKDF described:

> The vulnerability of these people was very high because everything they owned had been lost. So you know these are people who don't have bank balances or anything. They have their home. They have their home, they have crockery, they have cutlery, their assets. I remember this one lady in one of the villages, she said: "I don't have a sewing needle left—*Sooi bhi nahin reh gaee hai*," as you would say.

Understandably, the responses and perceptions of the affected community were also linked to the support, relief, and aid they received from the government and aid organizations. Perceptions of the effectiveness of the support received varied, but some common patterns emerged. The Pakistan army, typically affiliated with the government, was seen by most people to have played a crucial and important role in the early days as they had the manpower and transport to act and to reach people in need, despite the serious lack of rescue machinery. Samina Omar of Sungi Development Foundation in the NWFP pointed out that her organization would never have been able to reach some of the remote areas had it not been for the army with their sophisticated equipment, helicopters, and detailed maps.

Similarly, Ali Mujtaba, a volunteer in AJK, further noted the very important role of the army in providing access for relief and aid to reach the affected areas:

> The army's role was very, very positive when they restored to normal those areas where roads were blocked and there was no communication...I mean they cleared out the landslides and the roads, at least to the extent that they were accessible again, cars could come and go again, trucks carrying relief good could come and go.

As Tahira, a human rights activist, noted, there were some parts of the country only the army could access:

> I want to say good words about the army. The Pakistan army started helicopter drops to the furthest points of the Azad Kashmir mountains. The Kashmir mountains are inaccessible by road. And those were the worst affected by the earthquake because there were avalanches. You know, whole mountainsides came down, I mean the whole geography of that area has changed. There are lakes where there was no lake before. There is a new mountain where there was a lake before. So, there were whole communities stranded....There was no way to get to them except helicopter drops. Then there were some daring and dramatic rescues for people who needed to be hospitalised, you know. So there were helicopter rescues, there were helicopter drops of emergency food, emergency rations, emergency medicines, emergency tents, emergency bedding, emergency food, emergency medicines, emergency stuff, all air-dropped by helicopters by the Pakistan army. So I'd like to commend the role of the Pakistan army in the immediate aftermath.

However, the response of the overall government in the early days was considered by many to be inadequate bordering on negligent. Rizwan of Islamic Relief Pakistan, working in AJK, describes how locals and NGOs were forced to take on the immediate role of relief and rescue:

> In the first 2 or 3 days you didn't see any government official there. It was a situation of total chaos. In fact, these...er...BBC, CNN people, they would try and find someone from the government to talk to and when they couldn't find anyone, they would come to Islamic Relief and say, you are working here, you tell us...the army only started its relief and rescue after more than 3 days.

Strong views were expressed by everyone about the serious lack of rescue machinery and disaster management training, and the impact this had on the damage and loss that was caused. The huge loss of lives was particularly attributed to this. Tahira Abdullah, a human rights activist in Islamabad, commented:

> You can't do anything about the dead. Rescue the living. I am abso-lutely 500% convinced that had we had such training, had we had the proper equipment and trained personnel we would have been able to save at least one half of the 150,000 people who died. Unnecessary deaths. You know, the first lot that dies when the building falls on you...for everyone who falls...who dies when the building falls on you there are 5–10 who are alive and screaming to be saved but are not saved because a lot of rubble has fallen and we don't know how to remove that rubble.

Some level of organized aid started pouring in almost immediately, especially from the army and local NGOs. At the same time, many people expressed frustration that there were no cutters to cut people out from the building debris, that there was no other equipment, and that there was a lack of specialized knowledge about how to save people under the rubble, as journalist Abdur Rauf described:

> Close to Bagh a building had collapsed. We didn't have any machin-ery, any cutters in fact. You know we waited for 4, 5, days for you know teams to arrive from Russia and other countries. And some people were saved because of that. But a lot of people could have been saved.

In some cases, travel to communities was impossible. The govern-ment appealed for people to instead come down from the mountains for assistance in temporary tent camps.

As more and more aid agencies began to arrive and the distribution of relief goods and ongoing support began, a number of difficulties arose. While some areas were oversupplied, distribution was disor-ganized and sometimes inadequate in others. Food, water, proper rescue equipment, transport, communication, information, and edu-cation were insufficient in many places. Food in some regions was scarce due to the destruction of grain stores and other food stores caused by the earthquake but was quickly supplied compared to other needs, said local journalist, Abdur Rauf, in Azad Kashmir:

> I met the Director of the World Food Programme and I mean within 3 days they had gathered enough food and through us and other tele-vision networks also they wanted to air this appeal that we have plenty

of food, we don't need food now. So, I mean, this was a positive step that within 3 days they accumulated data and they knew what type of requirement people had as far as food was concerned.

Naila, then a teacher and a community worker in the NWFP, described how the lack of clean drinking water caused a serious health issue in her region:

> Dr. Farid Chaudhry from Abbotabad, he said that when the children were rescued from under the rubble of the schools 10, 12 days after the earthquake, many of them were still alive, but 70% of them died because they drank ordinary water that was being used in households. Those that drank mineral water survived, but those who drank ordinary water... at that time they didn't realise that the stream water they were drinking might have been polluted or something with chemicals.

In addition to its role in the rescue, the army was also made responsible for creating central camps for temporary shelter and relief distribution, called "tented villages" in various places, for distribution, and for managing the initial phase of recovery. With the one exception of Agha Murtaza, a volunteer from within the business community with a series of government, army, private funds, and corporate contacts at hand, everyone who had worked in the camps and had experience in working with the community expressed a strong view that stretching the army's role to include relief distribution and rehabilitation had not been a useful move, because they had no experience working with people who had experienced disaster and sometimes created more fear and conflict. Because they often did not know and did not have direct links with the communities, they often distributed assistance without checking what the needs were and who needed assistance.

The process of relief distribution was clearly problematic and stressful. Those who were directly involved in the process described how the people pushed, grabbed, and ambushed relief stations and helicopters, resulting in conflict and injury. Sungi Development Foundation executive director, Samina, described one village where a team of young men carrying sticks was created to keep order, and in other places behavior changed after it became evident that things had become unruly and posed a threat:

> In another one of the areas where we were working, one of our staff members got killed because of this. He was stopping people from rushing towards the helicopter. The helicopter had started moving and he was moving people away and his head went and hit that fan at the

back, you know the one that goes like this [moving fingers in circles],
and he died. So, yeah. Anyway, the people were really...felt really bad
about it in that area. And they were very good after that in organising
themselves better!

For others the frenzied grabbing of supplies did not tell the whole
story of how communities acted at that time. They described how
communities shared goods, organized themselves, and volunteered
in the process. Many people felt that this difference in behavior was
determined to a large extent by the arrival and the behavior of aid
agencies. As Ali Asghar Khan described in the NWFP province,

> I think the community was very together because they went through
> the whole tragedy together. So, I think that initial phase brought peo-
> ple together. I think the distribution probably tore them apart. As the
> insecurity mounted, in the hysterics of the moment, you probably you
> know grabbed for whatever you could get, but when you came back
> and you thought and you reanalysed your position and people around
> you, you know your family and friends and all that, so then you shared
> once again. And, so, because the social fabric was so well knit, I think
> interventions tried to damage it but couldn't.

Programme Director of OAKDF, Rashida Dohad, echoed this idea
that reports from the media and outside organizations coming to
work there were not always indicative of what actually occurred in
communities:

> A media person had come with us and was watching the whole pro-
> ceeding [distribution of relief] and he said: "You know I've covered
> a lot of this and everywhere *chheena jhapti* [snatching/ grabbing/
> pouncing] is what you see, and that didn't happen here." He hadn't
> seen that before. I think, a lot of people's reactions also depend on
> approach and no generalisations are possible. Different people reacted
> differently. Some were opportunists, but I think the opportunists were
> a very small minority.

One of the first challenges faced by the communities and the aid agencies was that of temporary shelter in the shape of "tented villages" or
centralized camps. Many communities were extremely reluctant to
live in centrally organized camps. While some local organizations
supported the rural communities in setting up temporary shelter in
or around their villages, the government and the army supported
by international NGOs pushed for and enforced central camps. To

Rashida and Ali of OAKDF, this reflected the government's lack of understanding of and importance given to the needs of the affected community:

> *Rashida*: ...and we even had of course Musharraf making bizarre statements like "Why don't those people come down from the mountains?" So, he obviously had no clue that if they come down from the mountains, what is going to happen to their lands, and if somebody encroaches their land, is the State going to protect it? Where do they keep their livestock if they come down from the mountains, etc , etc?
>
> *Ali*: They were insisting...
>
> *Rashida*: And if they come down from the mountains where will they go? So, that whole idea...that notion that when there is a disaster so the best thing for them, for everybody, is to come into camps and get fed and everything I thought was completely missing the reality of people's lives, yeah?
>
> *Ali*: Because logistically it might make more sense, but it's not going to happen, so why do you take a course which you know people are not going to respond to? And they kept on insisting "bring them down," "bring them down," but people were not willing to come down. So, we started setting up camps within villages, so smaller ones, 40 tents, 50 tents so that the villagers could live...they didn't want to leave their household goods which are buried over there, their land is over there.

Living in tented villages also rendered women further vulnerable to neglect and discrimination because of the strict *purdah* (veil or segregation requirement) that women in that area observe. Living in an open space where they would be exposed to so many nonfamily men if they stepped out of their tents meant that many were confined to their tents at all times.

Setting up tents close to their own village seemed to some local organizations, such as the OAKDF, to not only be more cost-effective, but also helpful in that it gave people peace of mind because they could live close to their homes and with their community, protect their assets, preserve a level of privacy, and help resettle their villages. There was also seen to be greater accountability in small communities, especially those that had been organized by local NGOs before the earthquake, because if some villagers were taking advantage of relief supplies, then the others would take them to task. The international aid agencies and the army, which supported the establishment of large camps made up of people from many different villages, differed from the local NGOs, which supported the people's desire to stay in their villages. Rashida further gave an example of how the OAKDF implemented the idea of tents in the villages, and of how she felt that the international organizations did not understand this despite their years of experience in disaster areas:

> The idea of the tents at a village, we used to call in a *saraai* [rest house], the idea of a temporary home and stuff like that, yeah? We saw other organisations pick that up. I don't know whether they had got the idea themselves or they were influenced by us. We don't want attribution, but it was nice to see that it was being followed, yeah? I don't think any large aid organisation followed it, but some other smaller organisations—national NGO-type people, who also had a sense of the local. I think they understood it better, but I think the larger organisations had too fixated ideas of how to respond and you know and of course, initially one used to feel a bit overwhelmed you know because they would come with 20 years of experience and then you would hear that they were specialists in disaster management, and one felt that "OK maybe they really know what they're talking about" [laughs]. And of course, I'm sure they brought a level of experience, but sometimes they were inappropriate.

International organizations, including the UN, were criticized by many for their "fixed" approach to dealing with relief and shelter, which was often competitive, disorganized, bureaucratic, and misinformed. Ali Mujtaba, a volunteer in AJK province, felt that aid achievement was often measured in terms of quantity rather than quality:

> It was like they were competing to win. So like, "we have reached this many beneficiaries" and "we have distributed our good to this many beneficiaries." And that was all quite strange because it was like they were working with their eyes closed. Someone who didn't need blankets would get like 4 blankets and someone who needed them would also get them. I mean without seeing whether someone else had already distributed those goods to the same people and if so it might be better to distribute them to someone else. If there had been a centralised system, maybe the situation would have been better. The situation would have improved faster. I mean even if we had worked under the one UN...all of us, all international NGOS had worked collectively under the UN, and if we had said, OK we'll complete whatever task the UN assigns, whichever region the UN allocates we'll go there. But no one did this. They all identified their own regions and started work there, whether or not someone else was already working there. And the result of this competition, of this sense of superiority was that, quite unnecessarily...I mean in some places there was full-on conflict—"no, these

are our beneficiaries so you can't go there," "you can't come here"—
things like that.

Others too observed that aid was often provided like a list of goods
being "ticked off" without sufficient monitoring of the process or the
impact. The sudden surge of money flowing from international aid
was also seen as being mismanaged, exploited, and almost "thrown
away" without any significant accountability. This meant that many
international organizations were offering community members and
volunteers significant and unprecedented amounts of money for
contributing toward the development of their own community. Ali
Mujtaba, who was initially a volunteer in AJK, and later worked as a
part of a psychosocial team in the NWFP, for example, was shocked
to hear the term "paid volunteers" being used by international organizations for short-term workers they employed in their community
projects.

The UN approach of managing the disaster through a "cluster"
system was also noted as a problematic process by many. While on
the surface it seemed that the UN was organized—it immediately set
up clusters of organizations for health, protection, education, livestock, food, water and sanitation, and so on, and taskforces within
it for gender, mental health; and so on—it was largely seen as a
disorganized, rigid, and bureaucratic system. A lack of coordination
and needs assessment before giving supplies, as well as of monitoring
and evaluation, were commonly reported. Many felt that there was a
significant amount of chaos, confusion, and corresponding delay due
to this cluster system, starting with a lack of clarity around which cluster was doing what, which cluster to approach, and how to delineate
what were seen as interlinked issues (e.g., mental health and health).
Many local organizations initially became involved with the clusters,
and some continued to stay involved in an effort to coordinate efforts
and participate in decision making, but eventually realized that it may
be more effective to adhere to their original approach of working with
and organizing their communities. Rashida of OAKDF described
this process as follows:

> We said, "OK let's try and influence policy," so we used to come from
> Abbotabad [NWFP], leave our on-the-ground work, but after a while,
> it was just running around in circles and we stopped. For example, the
> first few meetings they said, "Well you know the UN and the tsunami
> and this happened so the lesson from the tsunami is that we should
> have clusters, UN clusters." You know the UN has such a horrible time

coordinating so the whole approach to clusters to us seemed more of compartmentalisation. Our own approach...because we looked at the villagers, assessed what is needed and then you bring in whatever you need—housing, food, shelter, but the way they were doing it was the other way around. They were saying "how much food?—one cluster," "housing?—another cluster," and often they had no idea who was doing what. So, lack of coordination one could see would happen even when they talked and after you felt you couldn't have any money on the table, we stopped going to these things. We said: "Let's just get on with whatever we can do."

Another example of the mismatch between local and international approaches was the area of social development and psychosocial support. The UN, for example, was criticized by local organizations for not playing a more active role in promoting psychosocial support (or support through groups of people rather than one-on-one), pushing instead a clinical mental health approach focusing on one-on-one counseling, which was the government approach. An increased access to the remotest and most conservative areas for many meant that an opportunity to develop the communities further had opened up; yet they felt that this had not been taken advantage of by the government and the aid organizations, and insufficient attention was paid to issues of gender, education, and community development. The focus of the international organizations seemed to be on "giving and giving"material goods rather than on building and promoting long-term resiliency, sustainability, and development. Naila, then a teacher in a community school, felt that education should have been prioritized. Others too, particularly in AJK, felt that schools and universities should have been reestablished quickly so as not to interrupt education. This was considered so important in that region that Muzaffarabad universities requested an immediate temporary campus to continue students' education. Gender issues too, it was felt by many, should have been highlighted, but were not given enough importance. While gender was on the UN agenda and on that of many aid organizations, many felt that in practice aid agencies were blind to women's issues. For example, men were often the ones who were consulted, even when it came to the needs of the women in the community. Many felt that even toward the end of the emergency phase, gender issues remained and were not prioritized.

Local organizations, those that had local presence and were well-established before the earthquake of 2005, and the international organizations that worked through strong local partnerships had more success working with the communities to provide relief and shelter and focused more on development. A major reason for this was that local knowledge and participation in relief efforts were useful in responding to the situation quickly—for example, Rizwan Ashfaq stated in the case of Islamic Relief Pakistan that it was the only international NGO that had been working in Muzaffarabad and Bagh for over five years, so it had the resources and local familiarity with the people to begin work on responding to the earthquake immediately.

Similarly, Rashida and Ali of OAKDF noted that initially they tried to hand out relief goods the way some other organizations were doing, but soon made a conscious decision to go back to the way they had worked before the earthquake and focus on community-led development using village units:

> But that was the first and the last time we distributed goods in that manner, because you know human dignity was totally forgotten in that kind of a distribution. And then we said: "OK we're going to work through the organisations, we're going to empower them, we're going to, you know get them to do the surveys, get them to do the listing, we will only monitor, you know, on what basis the lists were made and whether the goods actually got to whom." So, that's how we decided to work.

Maria of Rozan also observed that the international organizations who worked closely with local organizations were much more successful than those that went in on their own:

> But local organisations who had stronger grounding in the community managed to maintain their own philosophy of working even despite this sort of immediate "we want to see action," "we want to see impact," "we want to evaluate what's happening," despite that pressure, they did maintain some level of, you know their own stance.

The most often and most vehemently discussed impact of international assistance was that it created dependency, a "beggar" or "passive" mentality that prevented communities from organizing themselves to respond to the situation and their needs. Community worker, then school teacher, Naila described the problem as follows:

> The many INGOs that were there—they created dependency, so people stopped doing anything themselves. I mean they [affected

people] were constantly chasing organisations that gave them flour, gave them lentils. They started depending on flour and lentils and where they would get supplies of blankets and sheets, and in this process, those very people who were initially helping each other became involved in who took more, who took less. It created conflict between them. So, a lot of the conflict was created by aid organisations because there was no planning around where and how fairly to distribute. They just kept giving and giving, which created a lot of problems later on. Organisations, such as Saibaan, who started work such as development and of involving community people, focussing less on relief goods and more on community participation, were overlooked.

Naila went on to make the point that the communities themselves were not to blame for being dependent, that this was a result of the way in which aid organizations operated in the affected areas:

> If organisations are going to go in and keep pumping in aid, obviously they would become dependent. If they [people] are dependent it is because they were made dependent.

The level of dependency created in the tented villages also meant that later it was difficult to resettle people back in their villages. Journalist Abdur Rauf noted:

> This is an easy life for them, living in tents, getting ready-made food. So, sending them back was another huge problem afterwards.

Others described the effects of living in camps where people had little control over their daily lives as weakening community and individual spirit. As activist Tahira said,

> The fear that those survivors experienced, the uncertainty. And then on top of it, you don't have a say in anything. Everything is being done for you, decided for you. It made people dependent, it made them weak.

How aid was distributed was seen as a significant determinant of dependency, as Maria, the codirector of Rozan, commented:

> There's a way in which you can do this [distribution of relief] that the community feels like yes it has suffered, but it has some level of control and it deserves or it has the right to the help it's getting, it's not necessarily charity. When we went in for a programme on psychosocial

support, for example, by that time the community was so used to getting things and going into a very passive mode, you know of being or not really believing that they have any agency or they need to take any action.

The provision of aid was seen to undermine the local work ethic. Rizwan, from Islamic Relief Pakistan, referring to the AJK area, gave an example:

One day we went into this camp and we met this woman, this girl, and she said to us, "You people have made our men completely useless." We asked her what we had done. She actually said, "It's because of you people that they [the men in the community] don't go to work anymore, they say that if they go to work, they'll miss out on the distribution that will be done while they're away." She literally asked us to stop distributing goods so that they would work!

Volunteer and later psychosocial support worker Ali Mujtaba said that local organizations used to try and encourage people to think outside this so-called dependency mentality to refocus on community efforts to work together, but this was a challenge:

We used to say to the people there, "Look—no one knows how long you're going to keep getting the help you're getting now, so what you need to do is revive the systems you had before. If there used to be a mud oven, we'll get flour from somewhere and you start making bread—after all, you had wood fires before too."

Aid also had consequences for groups of people other than people directly affected by the earthquake. The free flow of money used for paying volunteers and for dramatically raising the level of local salaries was also seen as having serious implications as agencies competed to be the highest-paying organization for "volunteers" and other workers, resulting in a general expectation of monetary remuneration for involvement in community activities. Similarly, staff from local NGOs were picked up by international NGOs at much higher salaries, and then let go two to three years later after recovery projects expired, as Rozan's codirector Maria argued:

We and other NGOs also raised at various forums with the UN that this is not ethical, you can't come in and start paying much, much higher salaries that are not competitive, and particularly with vulnerable groups because they were affected communities and needed the work, so their vulnerability was reinforced. So, yeah, that was an unfortunate dynamic of the whole thing. And even now, it's been what 4, 4 ½ years, almost 4 years and the repercussions are still there. A lot of people shifted NGOs.

Similarly, secure and well-built housing rents and transport costs went up as aid workers and organizations paid higher prices for these more scarce commodities, with stories of houses that once cost $150–350 per month in Ghazi Kot Township to rent become $1,500–3,500 per month, and that people were able to earn enough from driving one car to buy ten cars, leading to a rise in numbers of taxis and traffic on the road.

The sudden influx of aid funding also allowed smaller organizations to suddenly appear and gain temporary power. However, these organizations were donor driven and donor dependent, dissolving quickly as aid agencies finished their work and moved out. This, along with the tendency of international organizations to also withdraw as soon as their projects had been completed was seen as negatively impacting community development and sustainability.

The arrival and involvement of international organizations and their staff also created other dynamics. Given the fact that the earthquake-affected areas are some of the more conservative regions of Pakistan and are located close to those areas influenced heavily by the Taliban, aid organizations were seen by some extremist religious elements in the region as a Western threat. Thus, some of the existing anti-Western attitudes were strengthened with the arrival of international organizations, and several organizations faced threats of attack. Collaborating with local organizations and giving the local community more control, in some cases, helped relieve local fears and decreased the threat, as Maria commented about the NWFP:

> I think where they [international NGOs] partnered with local groups, I see more impact at the level of community. Where they went in directly as implementers I see little impact or less impact, primarily because it's less sustainable because many of them then withdrew... because there's a whole dynamic of Talabanisation or extremism that's going on in these very areas, there was a lot of backlash against foreign troops or foreign staff or foreign NGOs working in these communities. Perceptions in this area about foreign NGOs and even, unfortunately, local NGOs, because they became almost like one at this point in time, have become much more negative despite the fact that they have done a lot of good work.

These dynamics of the area combined with the overwhelming need of the time also allowed a number of conservative religious political organizations, including some banned ones, to strengthen their roots and flourish in the region, thus further reinforcing anti-Western attitudes, and in some cases reversing some aspects of community development, especially in relation to women. That there was no government monitoring of the organizations that worked in the affected areas and that there was a general lack of trust of government organizations was seen by journalist Abdul Rauf as a significant reason for why such organizations were able to influence people's thinking, such as

> that if anyone will get there to help them first it would be the *Jamaat-e-Dawa* [banned religious-political charity with links to militants] type of people. The trust they had established, that trust deficit is there as far as government agencies and other organisations is there. This trust has been established more by *Jamaat-e-Dawa*-type organisations.

Some ethnic political groups also gained popularity for their work at this time. The MQM (Mutthaida Qaumi Movement, an ethnic political party), for example, which has a strong influence in Karachi, Sindh, but not in the NWFP or Kashmir, was also seen as benefiting, winning the election in Muzaffarabad due largely to its visible involvement with relief after the earthquake. Similarly, community worker and teacher Naila felt that the government workers were also able to benefit politically:

> Even government people and heads, everyone was moving away from what they were supposed to do and thinking only of money. It was the same with our political leaders, they thought INGOs give more money, so they got involved there. There was a lot of politics there.

Despite the negative implications of the approaches and methodologies employed by many international organizations, many felt that their work had had a positive impact too. Where international organizations worked with the community and with local NGOs, supported local efforts and local capacity, and where they were flexible with their approach and funding, this was more likely, as Samina of Sungi described:

> Because we already had so much work in the affected places and all, so we agreed to it and said, "Yes do it here." I asked our board ... and

the reason we agreed was also our donors, one of our donors, the Norwegians, they rang up and said, "You know, whatever funds you have of ours you are free to use as you wish." So, we weren't tied down to what we had planned. So, that really gave us an opening. So, what happened was that I was comfortable saying that yes you can use the office as a secretariat. And then I didn't need to think about where the money would come from for whatever expenses this entailed. So, that was done. So, our donors were very helpful.

One level of empowerment that occurred, according to the executive director of OAKDF, Ali Asghar, was that despite politicians sending distributions to their voters to use it for political gain, their involvement opened up new opportunities for the community's political participation:

I think what was good that came out of this was it was a test for people…and we found that within villages political structures also changed in a sense that people who were seen to be unfair or unjust in distributing the goods that they have accessed or hoarding them, within the village itself were totally wiped out, meaning they had no standing. New people emerged in leadership positions within those structures, people who had been going out of their way to help people and all that.

The earthquake and assistance presented communities with an opportunity to change power relations and improve community life, as community worker Samina in NWFP province explained:

I mean we used to constantly tell them: "Take advantage of the situation." "We know that this area has always been neglected by the government, so, you know that this is the time that the whole world is focussing on you, so take advantage of it and make the best use of it, organise yourself so that more people will come, more organisations will come if they see that you're using what you want well." So, I think a lot of them understood that. That was very positive. And that's why we got reassured that bring organised and being prepared for such things is very important. Now, after this experience, what we have done is that wherever we're organising a committee, we're organising them also for preparedness for any disaster that comes in their community in the future.

Particularly for the rural areas and the NWFP, an effect of the earthquake assistance was that it "opened up" communities and people to the outside world and new ideas. Agha, a volunteer from the business community, shared his thinking about changed attitudes:

> Overall, I would say a lot has changed now because of the world media presence there and the world community and the world NGOs. It opened up a very new gateway for those people. They were very blocked-minded, very blocked-minded, they couldn't see beyond their noses. Now they've started thinking beyond their noses. They know there's a world out there as well.

Access to doctors, reproductive health, medical and health services in general, and education were some of the benefits of outside assistance, with many people reporting that awareness and community organization increased. Community worker Naila reported that some people talked about the earthquake as a blessing in the long term because it increased community organization and awareness to take charge of community issues:

> Even if they don't speak out or go protesting out on the streets, at least they discuss things amongst each other and have started to find their own solutions. What could be better than this? So, if a teacher can't go from Abbotabad to the mountains, they [community members] demand a substitute or if they ask the government to build schools, hospitals, provide health services. They've started to discuss all this with each other and believe me recently when it had snowed in those areas, this community woman asked how we were doing and all that, our project there had already finished, but we were still in touch over the phone. So, I asked how the roads there were and she said: "Oh *baitay* [literally translated as child, but an affectionate term used to address a much younger person] when you come, you'll fix them for us"! I said, "*Amma* [literally translated as mother, but a respectful term used to address an elderly woman], this is your job now, it's your road, your village, now you do something about it. What is your committee, the village based organisation that we had developed together doing about it"? She said: "*Baitay*, we'll do something, we'll find some way." Believe me, then the community people contacted their local council member and the mayor and the vice mayor and they cleaned and opened up the road. This had never happened before.

Some positive impact was also seen in the gender dynamics and attitudes of the community, as Ali Mujtaba's example illustrates:

> Amara [a female colleague] and I had gone to Besham to conduct a training with teachers and it was an all-men's group. And while we

were training them, we were discussing this—what changes have occurred in this area? So, one of the teachers said: "Look, isn't this the biggest change that we have a woman standing in front of us, running a training, that this is the first time in our region, in the NWFP, that a woman is training a men's group"! However, this change also had a backlash with community leaders accusing NGOs of promoting vulgarity and shamelessness of women and threats that a woman in an NGO car would be killed. But this particular change I saw more in the AJK side, not so much in the NWFP side because in the NWFP side they had always thought that these Americans who are coming, these white people, they're Jewish agents and they won't support us, they're here for their own agenda. So that created a lot of resistance. So, where there was a positive change, there was a negative one too. We saw both.

In some places, there were murders and attacks on aid workers. Naila Azam, then teacher, also spoke optimistically of the cultural change in women's role in the community after the earthquake and the aid that followed:

> Prior to this women could not have even imagined the way they benefitted from this. People's attitudes have changed. Now, education, what used to happen before was that they thought that education would allow girls to read letters, then they will elope with someone! So what is the point of educating them? But after this, women have jobs, they have teaching jobs, they have para-teaching jobs…they have a different role now, they've moved from a reproductive to a productive role. Secondly, women sitting and meeting—women-friendly spaces were made for women, village-level organisations were made, they were given the opportunity to participate in reconstruction and rehabilitation work. For example, Saibaan built this road. In our culture what role would women play in building a road?—but women monitored it, managed things, made the payments with their own hands, selected the labour themselves. And when such an excellent road was built they realised that women could be a great force for them, to develop the family.

Many organizations also arranged disaster-preparedness training, disaster information, and other capacity-building initiatives. The government and international organizations also improved housing in many areas, establishing safety standards. The earthquake also opened up opportunities for learning by organizations and individuals providing assistance. As mentioned above, community worker Ali Asghar, for example, explained how the organizations modified their distribution strategy after a chaotic experience in the community. Some level of modernization and development was also reported by some, but more in the smaller towns rather than the villages.

Despite some optimistic observations, many people remained pessimistic. As an example, some people compared the earthquake response to the current IDP situation unfolding in Pakistan as a result of the conflict with the Taliban in the NWFP and tribal areas, making the point that general approaches and mistakes had remained the same. Volunteer Ali Mujtaba also made a comparison with a smaller disaster in Quetta, saying,

> Well, there was a situation of sorts...a disaster in Quetta. And there for the first 10, 20 days people complained that no one was there for them. Then gradually people started responding and then reports started coming in that relief goods were being sold in markets, accusations of corruption, ERRA's (the government relief agency) inefficiencies...the same thing again. Again we saw organisations going there and doing their own thing and distributing things. There was no centralised system. The same mistakes were being repeated and we couldn't do anything. Even if we look at the IDP situation, everyone is doing their own thing. World Vision is doing their thing; the Red Cross is doing its own thing. Everyone is working in isolation. So, based on all this it seems we have learnt nothing.

Others too expressed similar views, especially related to the gender dynamics, arguing that women were still as vulnerable and would be in a similar situation if another disaster was to hit those areas again. Although some people felt that there was now an increased awareness of women's issues in a disaster situation, and that the UN was relatively more alert to it than it had been at the time of the earthquake, others felt that there had been no significant learning. As Tahira Abdullah observed in relation to the current internal refugee crisis,

> It's the same. It's the same. The government has not learned anything. In fact this time the women and the children, especially girls, are suffering even more than before. This time they're a deliberate target. You know what the Taliban did to girls' schools and to scores of women. And even in the camps...the women have been forgotten again. After the earthquake, we developed all these...manuals and booklets...you know that booklet that you guys [Rozan] developed on gender issues in a disaster with a whole set of recommendations? Is

anyone in the government reading that now? Has anyone bothered to learn from it? No!

Almost four years later, the 2005 earthquake that changed so many lives is remembered in different ways and cannot be forgotten. As Tahira Abdullah observed,

> The children who are growing up now four years into the earthquake have not forgotten. My little Danial comes back every year and lights candles where his brother fell down, where the apartment block fell down, the Margalla Towers. Every year he comes back from Karachi. He has not forgotten, I have not forgotten, nobody has forgotten. There's a huge crowd of people. I think we had 4,500 people lighting candles at the place where the Margalla Towers fell down last anniversary of the earthquake. So Islamabad has not forgotten. The Kashmiris have not forgotten.... The new city of Balakot is still not rebuilt. The government is locked in a battle over land acquisition because that old spot is...is right on top of the fault line.... The city of Muzaffarabad has been rebuilt but they have not forgotten. The children have not forgotten, those whose near and dear ones have died will never forget. It may be off the radar screen of the Government of Pakistan but who says the government has anything to say to anything? The people of Pakistan have not forgotten and that is important. The affectees themselves will never forget, those who volunteered their time and their effort and their money and their goods will never forget.

Apart from the traditional religious ceremonies around death anniversaries and visits to graves, some communities, organizations, and individuals mark the day with rallies, cricket matches, seminars, television reports, and reminders through text messages. Some people remember it as a blessing and feel that they have come out of it stronger and better. Yet others still express a feeling of despair. As volunteer, later psychosocial support worker, Ali Mujtaba commented,

> But when I went back to the same area after a year, they did not care. They say no one did anything for them. The people of NWFP would say, in fact, they still say that all relief goods went to Azad Kashmir. If you talk to people of Azad Kashmir, they say they didn't get anything and that it all went to the NWFP.

Such disputes about which group got what and disputes between aid organizations and local people were seen as contributing to conflict at the time. The fact that there are also places still not rebuilt and places where people are still living in tents is seen as reducing the peace of mind of the people and maintaining their level of frustration. OAKDF's Ali Asghar said that the remaining problems were unsettling for the people:

> If a child has to go to school and the school has not been built or if you find that the health facility is not built and there's no doctor over there so it is not as peaceful then. So, the reminders are there and that keeps on troubling you.

Some also talked about how the turbulent political, and law and order situation in Pakistan allows little room for people to remember and commemorate events such as the 2005 earthquake. Naila Azam who continued to work in the affected region notes, for example,

> Peace…as far as the earthquake is concerned, they've settled down, they're more at peace…markets are populated again, homes are populated again, work has resumed, they're earning their bread, their crops, agriculture, everything is happening, but attention has now been diverted to the current situation of the country. Now the things they talk about are political issues, peace, bomb blasts, the situation in Swat, etc.

Ali Asghar Khan of OAKDF, who is also still actively involved in work in the NWFP, reinforces Naila's view and described the dynamic changes that occur in the political life of Pakistan, thus affecting the ways in which communities remember:

> It's also about Pakistan you see. So much happens in this country that attention gets diverted to other things, like right now!

Conflict, Peace, community, and Disaster

Some people hit by the earthquake have again been affected by the conflict in NWFP, losing homes, livelihoods, and family members, and others experiencing physical violence. The conflict continuing in NWFP meant that things were not peaceful, and this had more of an influence on chances for peace than the earthquake, as explained by volunteer Agha:

> There's war going on over there right now. They're not at all peaceful. They were in a bad condition and they're coming back to a worse

situation because all the world's money that was spent on building schools over there, the Taliban have blown up those schools. So, they're at zero point again. The world community, if they really want to help, please come and build our schools, please come and give us text books, please come and build hospitals for us....And I want to convey a major thank you to all the people, the foreigners who did come and help.

Due to the increase in anti-West attacks, many NGOs have been restricted in the work they do in some areas of the NWFP, as community worker Naila said:

Non-government organisations are now becoming targets of terrorism. Local organisations or INGOs. They're specifically targeted. Recently one of our district officers was murdered, and another NGO worker was murdered.

This violence means that access of services and NGOs to the NWFP is reduced from the peak time of after the earthquake. This was often attributed to regional political differences, as volunteer Ali Mujtaba described:

In Azad Kashmir [AJK] there have been a lot of changes. Their government, unlike the NWFP government, which has completely isolated itself from foreign governments, stayed in touch with the governments of countries such as Turkey and Japan and they've developed full-fledged colonies, re-established their universities and all. So the situation in the NWFP is much worse due to the law and order situation and the war on terror. So, people's general perception is that all white people are bad. So, these NGOs themselves avoid going into the NWFP.

Azad Kashmir had more development after the earthquake, was more economically developed before the earthquake, and was generally perceived to be more liberal, open, and better-off than the poorer and more conservative, conflict-affected NWFP, so in that sense it was not seen as being more or less peaceful than before. One indicator of how the earthquake assistance perhaps created opportunities for peace was that in some cases restrictions on women and girls had lifted (e.g., girls going to school and women from different clans sitting down together), and so that contrib-uted to an opportunity for further cooperation and development in communities.

Box 3.1   Recommendations for Interventions

Many of the recommendations made in Aceh were also made in Pakistan, another largely Muslim state that experienced aid on a large scale from international and national agencies, following the disaster. However, people in northern Pakistan in rural areas specifically put emphasis on their reluctance to come to centrally organized camps, preferring to stay and protect their land and rebuild their homes. New recommendations from Pakistan include the following:

Support Local Organizations through Flexible Funding

- Allow the local organizations to make decisions about reallocating funding in an emergency. Trust and flexibility from donors can help local organizations act quickly to help disaster-affected communities.

Avoid Centralized Camps, Prioritize Rebuilding by Communities

- Decentralize camps to allow people to keep an eye on their property and rebuild their own lives. This also gives greater security for vulnerable members of the community such as women, who reported feeling more insecure in the camps.

REFERENCES

Ahmed, Akbar S. (1997), *Pakistan Society: Islam, Ethnicity and Leadership in South Asia*, Karachi: Oxford University Press.

Haleem, I. (June 2003), "Ethnic and Sectarian Violence and the Propensity towards Praetorianism in Pakistan," *Third World Quarterly*, Vol. 24, No. 3: 463–477.

Halvorson, S. J. (August 2003), " 'Placing' Health Risks in the Karakoram: Local Perceptions of Disease, Dependency, and Social Change in Northern Pakistan," *Mountain Research and Development*, Vol. 23, No. 3: 271–277.

Islam, N. (2001), "Democracy and Governance in Pakistan's Fragmented Society," *International Journal of Public Administration*, Vol. 24, No. 12: 1335–1355.

Malik, Iftikhar, H. (July 1996), "The State and Civil Society in Pakistan: From Crisis to Crisis," *Asian Survey*, Vol. 36, No. 7: 673–690.

Malik, Iftikhar, H. (1997), *State and Civil Society in Pakistan: Politics of Authority, Ideology and Ethnicity*, London: Pan Macmillan Press.

Mumtaz, S. and Shaheed, F. (1987), *Women of Pakistan: Two Steps Forward, One Step Back?*, Lahore: Vanguard Books Ltd.

United Nations (2005), *Pakistan 2005 Earthquake, Early Recovery Framework (With Preliminary Costs of Proposed Interventions)*, Islamabad: Pakistan.

United Nations Development Programme (2002, 2007 editions), *Human Development Report*, New York: UNDP.

4

Solomon Islands

Anouk Ride and Diane Bretherton

Anouk Ride is a resident in Honiara, Solomon Islands, and traveled to the Western Province to research this chapter. She has previously conducted interview research in Solomon Islands, Bougainville, and West Papua. Her works on peace and conflict, environmental conflicts, communication and indigenous cultures have been published in newspapers and magazines in Australia, UK, and the United States, and her most recent nonfiction book The Grand Experiment *(Ride, 2007) has been nominated for an award.*

Katrina Parkinson, from Gizo, provided valuable liaison and translation assistance for this research.

I thought trouble is coming, we are going to face big trouble and not very long after that we heard this sound more or less like trouble, a whistling sound, before the water comes up sheesh, sheesh, like that. Very big sound. There was the sucking of the water down, everything just sucked down, you know, like the sound of a waterfall, and then all of a sudden the sea came up, big water, it didn't come up like a wave, it was like high tide.

I called out to the communities, the small groups of families and told them to go up, I shouted and shouted and shouted and they tried to escape but the water was so fast I saw everything. I saw everything, the water catching them, but we couldn't make it, we just couldn't make it. It was so strong and so fast.

Neemia Boberio, community coordinator, from Titiana Camp 2

Twenty or thirty minutes after the earthquake I made the first call, the normal line was washed out and didn't work, same with the mobile, we just had the satellite phone. Every store had water inside, all the food was wet, every service was closed, everyone just left. The hospital by the sea had no evacuation procedures and we just took people from the hospital up the hill in the truck.

Everyone went up to higher ground and started helping each other, organizing themselves, "We'll stay here," "We'll do this," no one told them what to do at that time. Later they came to us and said: "This is what we want, we really need tents."

Sipuru Rove from Kolombangara

CONTEXT

Before the tsunami of 2007, reefs such as those in the Western Province of Solomon Islands were recorded to have one of the highest number of fish and coral species in the world and attracted tourists, in particular divers. In this province, canoes and outboard motorboats are the most often used form of transport, and fishing is vital to many villages.

Solomon Islands, so-named after a Spanish explorer who found gold there in the 1500s, is a string of 992 mountainous forested and flat coral atoll islands, a third of which are uninhabited, bordering other island states of Papua New Guinea, Vanuatu, and Fiji.

Sitting on the Pacific "Ring of Fire," Solomon Islands has four active volcanoes, and its location (on latitude 9 degrees South, 160 degrees East) makes it conducive for tropical cyclones, with five cyclones experienced in 2003. Many indigenous communities have stories or widely known strategies that suggest experience over time with disasters, with the rainy season each year from November to March also familiarizing people with heavy rains, storms, rough seas, and seasonal flooding. Records since 1926 suggest tsunamis have been experienced in the Western Province area every decade except for the 1940s and the 1980s (NDC 2007: 59).

Land in the Solomon Islands is held predominantly by traditional owners. When the British colonial government took administration of these lands, they began to develop flat lands suitable for agriculture and development on the coast. Between 1886 and 1920, the colonial government and the traders had acquired possession of 22,720 acres of land on the plains of Guadalcanal, the most fertile and easy to farm land in the Solomons. Development of palm oil plantations and timber logging continued in earnest, as Solomon Islands became part of the British empire in 1893 (Naitoro, 2000: 5).

The sea and the land also provide the basic needs of the population, with 80 percent of the population having a subsistence lifestyle through farming, fishing, hunting, and gathering. Being rich in the Solomon Islands means owning Western consumer items and assets such as cars or motorized boats for transport and permanent housing (rather than wood and leaf-roof housing). As surgery facilities for brain, heart, and other internal organs are unavailable in the country's main hospital, only those who can afford overseas travel survive complicated health conditions. Almost all ownership of property and businesses in towns and cities are run on a communal basis by families and extended family networks, and in the rural areas collective traditional ownership is the norm.

Solomon Islands ranks number 129 on the United Nations Development Programme's (UNDP) Human Development Index (UNDP, 2008: 231). Subsistence lifestyles and rich farming and fishing areas means that the population has a low rate of underweight infants at birth and malnutrition, a life expectancy of 62 years and a mortality rate of 29 per 10,000 births for children under five years. However, it also has the highest malaria rate outside Africa, and little data exists about other infection and disease rates, including HIV/AIDS.

Prior to the 2007 tsunami, Solomon Islands had US$262 per capita in aid a year (UNDP, 2006: 345). The biggest expansion of aid and external assistance, much of it dedicated to law and order and governance programs, occurred after 2003 and the arrival of the Regional Assistance Mission to Solomon Islands (in 1998 the rate of aid was US$115 per capita).

However, despite this influx of aid and development of tuna and timber exports, there are projected problems with further economic development. Estimates in 1998 were that viable forest for logging will be gone in 10–15 years (World Trade Organization, 1998), and some species of tuna, the other major export, are heavily fished. Land tenure uncertainties and conflicts have also stymied further businesses in Solomon Islands, especially on Guadalcanal where mines and plantations have often been a focus of protest, strikes, and sabotage.

Although Solomon Islands is classified as a Melanesian country, this can underplay the diversity of the country that is predominated by a mix of Melanesian tribal communities plus significant Polynesian and Micronesian communities, which contain both historic and recent settlers from other parts of the Pacific Islands. Cultural norms, arts, religion, traditional money and exchange, leadership systems, and communication styles of people can differ substantially from one group of islands to the next. There are around 70 indigenous languages and 100 dialects spoken in the Solomon Islands, with *pijin* being a lingua franca for different language groups and English the official language. Solomon Islands *pijin* uses a blend of words from local languages, English (often terms from colonial days), in a continually evolving language, often using acronyms and slangs for new concepts and technologies.

One of the commonalities to the different communities of the Solomon Islands is the strong role of religion—both Christianity and indigenous belief systems. Before the arrival of missionaries, spiritualism in the Solomon Islands was dominated by local beliefs in magic, sorcery, shamanism, taboos, and beliefs about the afterlife (Alasia and Laracy, 1989: 83).

Missionaries soon established large Catholic and Anglican Church networks. Today, 95 percent Solomon Islanders identify as Christians with 34 percent Anglican Church of Melanesia, 19 percent Roman Catholic, 17 percent South Sea Evangelical Church, 11 percent Uniting Church, 10 percent Seventh Day Adventists—minority faiths including Baha'i and Islam. Churches often are important providers of services and social support in communities and play an important leadership role in key events such as mediation of conflict (Alasia and Laracy, 1989: 84). Many people follow indigenous beliefs and Christian religious rites for spiritual and practical needs such as in attempts to heal the sick, death rites, and marriage ceremonies.

Accepted norms of behavior specific to groups of people, often rooted in tradition or old ways of doing things, known as *kastom* (custom) is a strongly valued and publicly supported part of Solomon Islands life. *Kastom* can be used to refer to types of dancing, singing, stories, art, fishing, farming, and rules for relations between men and women and elders and youth. Often outsiders have equated *kastom* and tradition as the same, however, many precolonial traditions changed, adapted, or were removed as a result of interaction with other Pacific Islanders and colonialism, so *kastom* is an evolving rather than a fixed entity. As historian Clive Moore explains, *kastom* "is also very abstract and appeals because of the diverse construction that can be applied to it by both the most modern 'educated' and 'sophisticated' citizens of Melanesia, and the 'grassroots' [traditional rural] people living in isolated mountain villages. *Kastom* has been the symbol of cultural autonomy, and of resistance to cultural, economic, or religious subjugation" (Moore, 2004: 27).

Another key commonality across the islands is the strong familial and social networks of reciprocity and obligation. An individual's daily life is determined by obligations to several layers of family, kin, and social networks. This begins with the inner circle of immediate family, then broadens to the extended family (laen or tribe, kin networks, sometimes including peoples believed to be from the same ancestor/s)—and then further broadens to fellow members of their local community and language. The latter is often referred to as "*wantok*" (one-talk), although *wantok* is used generally to refer to anyone with a familial or community connection with which a person has a shared social origins and obligations. Church groups and activities are also a key social network and organizational hub. These social ties can take precedence over other newer ties and bonds.

This strong and enduring social system in Solomon Islands often clashes with introduced Western concepts. For example, local politicians are beset with obligations to *wantoks* once elected, formal employment and "getting ahead at work" is often secondary to supporting other *wantoks* in the workplace, and in most cases avoidance of conflict is preferable to disrupting *wantok* links or their interests. Instead, most people tend to rely on the community first and Western institutions as a last resort—for example, most people prefer to resolve a dispute with a neighbor through a chief or customary law rather than courts (ANU Enterprise, 2007: 10).

Given this low level of support for government and reports of corruption, Solomon Islands is often classified as a "failed state" or "fragile state"; however, the *wantok* and *kastom* system provide many functions of the Western state and has led others to classify the country as a "hybrid political order engaged in a complex process of state formation, albeit with the additional dynamics of external assistance, and rather poor articulation of state and custom" (Boege et al., 2008: 20).

Society and government are in a process of recovery following violent conflict, colloquially referred to as "the tensions," between militant groups from Guadalcanal and Malaita. Often labeled "ethnic tensions," they had roots in pent-up issues surrounding distribution of economic and political benefits from national government and businesses, both of which are concentrated in the capital of Honiara on Guadalcanal island, where migrants from populous Malaita form the bulk of the workforce. While Guadalcanal was the site of the most deadly conflict, the Western Province also experienced an increase in tension between settlers from the island of Malaita and indigenous communities, and, at the same time, the presence of former militants from the Bougainville conflict over the national border also led to fear and some incidences of violence.

In June 2003, the Pacific Islands Forum agreed to establish the Regional Assistance Mission to the Solomon Islands (RAMSI)—this is largely staffed by Australian government and defense forces, so locally seen as "the Australian intervention," and also has New Zealand and Pacific Island staff playing key roles—under the supervision of the forum. Originally RAMSI's focus was disarmament and basic peace-keeping, but it has now broadened to "state-building," with a focus on law and order.

However, after the successful disarmament phase and relative absence of violence, in 2006 riots that involved targeting of Chinese businesses, following the election of the prime minister, indicated underlying problems. These include conflicts relating to disputes at the oil palm plantation and gold mines on Guadalcanal, demand for alienated land to be reverted to original landowners from around the country, population growth leading to pollution, overcrowding and stress on health and schools, plus high rates of domestic violence and the proliferation of squatters, and unemployed youths in Honiara and Gizo.

Many buildings in Solomon Islands' second-biggest urban center, the town of Gizo, located on the island of Ghizo, were washed away when at 7.39 A.M., on April 2, 2007, an earthquake measuring 8.1 on the Richter scale at its epicenter, 45 kilometers southeast of Ghizo Island, Western Province, caused a tsunami, with waves of up to 10 meters, that affected islands of Simbo, Ranongga, New Georgia, Mono, Vella Lavella, and Kolombangara. More than 20 aftershocks (5–6.8 on the Richter scale) were felt in the coming days, and in some parts of the province, land was left exposed. Ranoggah island actually was lifted three meters, and the extra land above sea level exposed reefs, mangroves, and coastlines (NDC, 2007: 10–11).

An estimated 11,322 people were affected by the tsunami in the Western and the Choiseul provinces (17 percent of the population). Many of those 52 people who had died had been at sea or close to the coast when the tsunami struck.

Gizo's nearest airport, Nusatupe, was closed, government agencies had no telephones in Gizo, transport around the area was largely brought to a halt, and the radio station was badly damaged and unable to operate. All of the Chinese stores in town are close to the sea—stock was damaged and the stores closed. Around 200 schools were damaged, and the two main hospitals in Gizo and Sasamunga were badly affected (UNICEF).

Landslides led to loss of farming land with blockages of rivers and paths with the debris further holding up transport. Supplies took many days to get to the Western Province due to rough seas and lack of a functional airstrip close to the affected areas. (NDC, 2007: 7). This led to food shortages in the province, particularly in and around Gizo town.

RESEARCH OUTLINE

Stories of this disaster were compiled through a total of 15 interviews conducted with the set of structured questions supplied by the researcher, who is from Australia and a resident in Solomon Islands. Interviews were conducted in and around the town of Gizo, Western Province, in June 2009. All interviews were done in *pijin* with people choosing this language as preferred over English, which many people also speak. Interviews generally lasted one hour, and were more of a "storytelling" style following cultural norms that tend to take a narrative approach to sharing information, sometimes interviews stretched to two hours to allow for storytelling and further discussion. All interviewees said that this was the first time that they had been asked to reflect on the experience of the community regarding the tsunami. People from areas affected who were interviewed include the Gilbertese community of Titiana (Camp 1 and Camp 2), the Malaitan fishing village, Gizo town, and the close-by island of Kolombangara. Five women and ten men who played key roles in the community response to the disaster were interviewed after being approached with the aid of a local guide and referrals from the NGO workers and other community members. Most interviewees chose to have their names included along with their comments, but a few opted to prefer to remain anonymous, often citing the smallness of the community and the need to not cause offense to maintain good working and community relations as the reason for this decision.

COMMUNITY RESPONSES

As the tsunami struck early in the morning, it took people by surprise, first there was a large earthquake so strong that some people could not stand up. Then there was a loud whistling wind sound as the sea water was sucked back from the shore way out, exposing the reefs. When the sea came back, it came in waves that did not look high but had an enormous strength, the waves moved as fast as a car, and similar to a car knocked down everything or everyone in its path including coconut palms and large logs.

This was the first, big natural disaster that almost all people in Gizo had experienced in their lifetime. Some had only just woken up. Some people even ran outside naked, such was their shock and alarm at the time. People ran up hills, and some who had no hills scrambled up trees as the water approached.

Everyone either experienced or saw other people they knew experience some kind of stress, trauma, or grief at this time. Some people could not speak or move for a time after the tsunami. People were afraid to be alone, afraid to go out from places where they gathered on high ground in search of food, and afraid of the sea. Children in particular were described as frightened and distressed, often their behavior being more difficult to manage after the tsunami. There was a lack of official direction or coordination about what to do in the days following the tsunami, and the first response was by the communities themselves.

Some people said, looking back on it, there were signs something about the weather was unusual. Neemia Boberio, community coordinator, from Titiana Camp 2 said people had noticed strange things:

> People told me that during that Sunday night there was early signs— they saw the color of the moon change and the behavior of the dogs was a little bit strange. When the earthquake shook and just before the water came up, the dogs ran away. I didn't know then but if I just had told people to run then we wouldn't have had any deaths, we had 3 deaths. We didn't do the right thing by running up the hill, the hills were very close to us, just next to us, not far from us. But we didn't make it to the hills because it was the first time this happened. We were in shock. We were standing just watching the sea.

Other people also said that they had knowledge that helped them understand that a tsunami was happening, with many people mentioning films about Indonesia and Papua New Guinea tsunamis as helping them understand what the signs were that a tsunami could happen.

Many people said it was lucky that the tsunami struck early in the morning before people had a chance to go out from their houses to school, work, or fishing. Because they were at home, they could quickly gather their family together and move to areas of safety (such as higher ground) when the earthquake struck, so their lives were saved when the tsunami waves came. But in Titiana, many people were outside when the tsunami happened, and so more people were killed or injured. A farewell to a visiting church delegation and the opening of a kindergarten happening in Titiana that day meant that more people were awake and down by the sea.

Everyone agreed that in the first instance, people cooperated to help each other survive, such as telling people to run to higher ground, helping people who needed to be carried or helped to move to higher ground, rescuing some of the people who were trapped (particularly the people who were older, not fit, sick, or the children). Often people made sure their own family was safe, and then helped others such as groups of children and people at the hospital.

Then, people cooperated to respond to the situation. They identified the missing and reunited separated people with their families. In Titiana, one of the immediate responses was to find their dead and bury them. Everywhere, people took care of the children who were the first to be given food. Local Chinese business owners were the first to supply food on a large scale (at Ja Mountain near Gizo and Titiana), which they did by sending teams of boys and men from places where people gathered on higher ground down the hills to the washed-out stores to find food and drinking water supplies and bring them back to share.

People also sat together and told stories, sang, or prayed to ease their minds. Everyone reported that while people were in shock and were sad, the general response of the people was to help each other at this time, and the levels of help people gave each other was high.

Achievements of communities at this time include donations of bush materials from some islands to other islands to help people rebuild houses, donations of food and clothing from those who were not badly affected to those who were badly affected. Volunteers contributed to relief efforts, including around 120 Red Cross volunteers alone. People from the Kastom Gaden Association also assisted people with supplies and ways of planting crops, especially those who had not done so before because they previously relied on fishing for income to buy food.

Churches and women's groups had their own contacts, they called representatives in the communities they assisted who handled the distribution of assistance. Other places such as Titiana created village committees to decide what to do, as Everlyn Mose from Titiana 2 Camp explained:

> Not all of Titiana came up to this place so we made a new committee, for looking for a place to settle, tents, food. We held a meeting to decide what men, what women would be on it. When organizations [from outside] came, that way, we knew very well what needed doing.

A group of women from different churches held a meeting to decide what to do a week after the tsunami, organized the collection and distribution of donations of soap, clothes, and food. Churches brought people together with most churches holding services about a week after the tsunami, and some such as in Titiana reporting more people went to the church after the tsunami than before. Many people mentioned that achievements during this time could also be spiritual and emotional:

> An achievement was the joy we had from touching people's hearts. (Miriam Maiice, local women's group coordinator)
>
> A strength of us, despite the problems we were facing, was faith, it lifted us I tell you, it gave us peace. When we prayed, visited other families, prayed with them, worked with them and organized donations, it lifted our spirits. (Moffat Maeta, church member)

People mentioned that the tsunami fostered a communal spirit:

> It changed our life, before people had things they might not share but now you don't know when everything you own might disappear, so what's going to happen if you don't share now and then you need help? That changed inside the community. (Oscar Titau, church leader)

However, after this emergency phase, most people agreed that cooperation of people decreased in and between communities around Gizo. Everyone interviewed mentioned that land or distribution of aid and assistance or both being the big causes of decreasing cooperation and in some cases increasing arguments and other signs of conflict in communities.

For people in Titiana Camp 2, a community largely made up of people from Kiribati (which experienced large-scale migration due to food shortages and sea-level rise), their land status in Solomon Islands is insecure. This was a particular issue that they felt was holding up development including building of schools, permanent water supply, houses:

> We talked with Save the Children [about establishing a school] but we did not revive anything yet, just the activities after the tsunami which were good to take their minds off things. I think we need a kindergarten but with the land insecure we can't do it. But education of children is important regardless of land. (Everlyn Mose, from Titiana Camp 2)
>
> We raised many concerns, some of which were addressed because they did not pose any problems for aid agencies, but when our needs has something to do with land, nothing. These agencies love to help but

when it comes to do with anything involving land, forget it. (Neemia Boberio, community coordinator)

Signs that people pointed to as demonstrating a lack of cooperation were the rate of people volunteering to help at school activities, and initial numbers of volunteers such as in the women's groups decreasing after the emergency time passed, and other family commitments took precedence over volunteering.

Some people said that those who had experienced natural disasters before were more able to do things after the tsunami and had less shock. For some communities, the level of cooperation decreasing after the emergency phase was not too serious in its impacts as people went back to their families to rebuild again and resume a normal life. Some people mentioned that villages where leadership was strong were responding very well to cooperate and rebuild. For instance, a church volunteer commented,

Being able to share and care for each other was a strength in this time but we should not forget it and build up on it over time. One village of a relative is a good example of that—if someone needed to build a house everyone would go help, they were always doing everything together. The chief would ask people: "What do we want to do?" People would then decide and do it, it's a nice well kept village, everyone says so, I think that old man brought people together after the tsunami.

Others mentioned the role of culture, particularly in the villages, in community response:

I think during that time a nice thing was it brought us back to culture—we have a chief as our leader and if anything happens we listen to him but we forgot that, this tsunami brought these ways back. Any call to action the chief did, people responded, not like before. I think that's a good aspect of people's response. (Kastom Gaden worker and trainer, Silas Kere)

For other communities, as cooperation decreased, conflicts began or started up again. Problems that had always been there became more dramatic or more serious because of the tsunami. The expectation of assistance affected how people responded to the disaster:

Many people struggled, especially the time the tsunami came, people were stressed, but after one or two days everyone got up and started

doing something. On the negative side, a weakness maybe is that people were expecting too much from people from outside who came to
give help. I went around and saw people with broken tents six months
after the tsunami and I said: "What are you doing? Why don't you get
up and work to help yourselves, do something now?" Most people it
took them two days of rest because of the shock and then they started
to rebuild. It was better to finish a house with local material than to sit
around waiting. (Religious leader)

When it comes to work, it's a bit hard for getting cooperation. When
it comes to the blessing of supplies, the sharing of the food, you will
see the whole place here. When people come out from the aid agencies they must come with something to lure the people out [laughs]
yeah, bait them out from their small holes. (Neemia, community
coordinator)

Common things people needed but could not get were food, safe
drinking water, and shelter—more often harder to find in urban
areas than in rural areas where traditional shelter (sago palm) and
some food sources (such as cassava, taro) were not destroyed by the
tsunami. (Titiana was described as an area particularly resource poor
and badly affected with whole villages wiped away, whereas other
places had more resources and less damage.) Also mentioned as things
needed were medicines, school or kindergarten or activities for children, clothing, counseling, church halls, and help for businesses to
start up again. The first food supplies generally arrived two or three
days after the tsunami in Gizo, but some communities, for example,
in Kolombangara mentioned not getting assistance till two weeks
later. Tents were in short supply.

Money was also something some people needed to start rebuilding
houses and replacing boats. In the case of the Malaitan fishing village
money provided was not seen as enough:

We had many houses gone and there was just $1,000 [Solomon dol-
lars, or approx US$100] per family for building houses, I spoke out
on the radio and said: "How do you build a house for $1,000? Maybe
you can build a house for chickens but not human beings!" (Malaitan
chief, Lawrence Mana)

Many people mentioned that it was difficult to figure out what the
different NGOs were doing at that time. Local journalist Adrian
Ginia described how the radio station became a way for people to
ask questions about aid and to talk about problems, frustrations, and

rumors they had after the radio station was restored and functioning in the third week of May:

> Straight after we started all the NGOs started using our program for awareness programs about health and all these things and we had news coming in from people about supplies and arguments. The main issues [people had] were distribution, some things didn't make it to their destination, they got lost along the line. Even the NGO workers, they ended up with the most benefit from distribution without assistance reaching those who were really victims. All these things were said [on the radio] they came in different forms but they were complaints about distribution of supplies. Others said "why not us"? "Why is World Vision doing something in this community and not our community?" On the other side, the Red Cross, Oxfam et cetera have their own stages of the process—emergency relief, rehabilitation and reconstruction and they tried to make people understand.

From the provincial government perspective, they pointed out that while there was a central coordination point for assistance through United Nations Disaster Assessment and Coordination (UNDAC) and then the provincial government for assistance, not all the NGOs worked under this system, and so the local authorities were unable to keep a track on what assistance was being provided. Activities funded by the church generally fell outside of any coordinated government efforts.

At the same time, rumors and radio reports about the large amount of assistance coming affected people's expectations, particularly when it was announced that the local Members of Parliament (MPs) would be given a total of (Solomon dollars) Sl$15 million (approx US$1.5 million) to distribute to their constituents as they saw fit. Originally, the National Disaster Council was to carry out a shelter policy, approved by the cabinet, that endorsed the provision of help with house frames and then cash to buy local timber and palm to complete the houses—Sl$15 million was allocated with aid agencies to kick in Sl$25 million. However, when the Cabinet opted to give the money to the MPs, the aid agencies withdrew funding from the project (instead NZAID and AUSAID gave the money to their national NGOs). One local NGO worker explained that this idea of large sums of money coming put the NGOs under pressure:

> Sharing was hard for people, their immediate concern was their own family, and to make it worse, people heard there were millions of

dollars of assistance pouring in so people were asking: "Where's the money?" "Give the money." It was complicated. People were cross, people were fed up.

For local NGO workers, the lack of cooperation between NGOs meant that some gave up with sector meetings:

> The aid organizations all took an area each, Oxfam Gizo, another organization Choiseul etc. where they would work. And there was a lot of flag raising by the NGOs, it was like a competition, who is doing the best? So, the coordination meeting was not interesting. We should have been talking about how to improve things, instead you saw one organization belittling the other one, I thought: "This isn't helpful." We didn't want to go anymore. We just thought: "Forget this meeting and continue our operations on the islands." (Local NGO worker)

Confusion between aid agencies also had a doubling-up effect with one local church volunteer commenting:

> Each different organization had their own way of dealing with the disaster. At that time no one really thought of working together but in retrospect that's what should have happened because the same people were getting the same things and you'd hear stories about people having so many pots, so many things and other people were getting nothing. The government too: all the money that they were getting for the tsunami, where did it go?

Local villages were bewildered by the lack of assessment of some organizations:

> *Anouk*: Did the assistance support community achievements?
> *Silas*: I think that's a good question, some come and support us and some just come in the name of distribution and that's it, but it's not right, some organizations don't go by the local village community there or even come in and talk to the villagers, some just come along the seaside and throw things on the wharf and rush off to finish their business. They don't come to monitor what was given previously, some came one after the other without much thought, maybe they were bringing the same things. I think more important is to come assess the situation too.

People thought that it was the people who asked for assistance who got it, whereas the people who did not ask did not get it. Those who knew the aid workers were also more likely to know about it and get assistance:

> For example chainsaws, there's one person working for an international NGO and all the chainsaws end up in his family and other communities know that it is only he who has the 10 chainsaws and has benefited from them when it should have gone to the whole community. So, all of these conflicts happened everywhere. (Adrian, local journalist)

Church volunteers also mentioned the chainsaws as an example of unfair distribution, and the local Provincial Secretary Arnold Moveni pointed out that this was a benefit that outlasted the rebuilding phase as people could now use them for purposes other than building back houses after the tsunami. Apart from chainsaws, tools, kitchen supplies, and water tanks were other high-value items that some people had more of than before.

Unfair allocation of assistance led to frustration:

> I had staff who lost their houses, our house was also damaged, I asked for 4 tents from whoever was sharing and giving out tents but I was not able to get anything at all. They said to us: "You work with Chinese people, they've got money, they should look after you." I mean that's nonsense. Because at that time, who can rebuild your shelter straight away? And there was an immediate need for shelter but we couldn't get it for any of our staff and I think that's bad. We tried all of the groups, except when I came back I went to one, a group called Shelterbox, they had these boxes but they were short of them so they gave me one and we were waiting for 2 or 3 more weeks and the boys they just couldn't bear living under the canvas, some got threatening they said: "Look, if you don't give me a tent now for the family we are going to burn the others." So, then they started giving them tents. (Local shop owner)

The Provincial Secretary Arnold Moveni pointed out that while some village disaster committee leaders distributed resources fairly, others only distributed it to their families, and this experience then put greater need for the officers to go into communities and properly monitor the distribution of assistance. Frustration with the distribution or lack of distribution by the MPs of the funds allocated for the tsunami relief was common, with many saying that till today they do not know where this money went.

In other cases, assistance was seen as having strings or being impractical, such as limits on what community tents could be used for:

> We needed tents and they gave us one that fit a family only, the big tents they said were only for use for temporary schools. When I went back to the province well I thought it was stupid so I asked a stupid question I said: "Look, does the tsunami come and only affect schools or does it affect everyone?" But they said the big tent was only for schools. (Chief Mana)

Local people working in local organizations could be disappointed with their national colleagues because of the lack of consultation and understanding of the environment that the national staff had about where they were providing assistance:

> Proposals people make for building, maybe those people have only lived in Honiara. They didn't consider the costs of not just cutting timber but transportation, in the budget, they did not know what was involved in building here. They must understand time, distance, the environment but instead the project stops half way and has no more budget because they don't have any experience working here. (Moffat, church member)

Some people became disappointed that promises of assistance were not met—with many in Titiana believing Oxfam promised houses, but then there was nothing done because of land insecurity.[1]

Assistance that the local services had been seeking for some time came in the wake of the tsunami—a school complaining about its roof for ten years finally had it fixed, and the Gizo hospital, which services the whole province and was previously not considered a funding priority, is now being allocated funds.

However, some people pointed to behavioral changes that they attributed to the experience of disaster and tsunami assistance. One person said that the provision by the aid agencies of free condoms and payment for youth volunteers resulted in an increase of drinking and sexual behavior of young people immediately after the tsunami. Other people had seen people with cash payments designed to help people build houses spending that money on other things.

While some people thought that in the case of an emergency like the tsunami it was all right to take goods from stores, other people were disappointed with this and called it theft, pointing out that many items taken were nonessential such as cigarettes and alcohol.

Reports were made to police, but they were unable to gather evidence and prosecute.

Oscar, church leader, Titiana Camp 1, pointed to Oxfam's assistance in donating a boat to Titiana was useful as was some other organizations' assistance with planting vegetable gardens around Titiana. However, most people said that aid did not support the ways that people had of making a living:

> Some people get more, plus, plus, on top. Some individuals ended up with plus plus, some families getting more than one shelter for instance. But none of the businesses got any help from the groups. For me I learned we have to be more careful with stock, insurance does not pay for damage to everything. (Local shop owner)

Many people said that the tsunami was in some ways a blessing because it brought assistance and development to Gizo that otherwise would not have happened. Some of these blessings mentioned include road building, communications infrastructure, a new TV channel, greater resources in emergency preparedness and systems, the idea that a town council would be useful, children's activities, assistance for fishing including supplying boats lost in the tsunami, wharf repairs, and school repairs.

At the same time, environmental changes and assistance had an impact on local business, particularly reliant on tourism:

> Another change here and some other places in Western Province was that the tsunami destroyed the reefs that provide livelihoods for people. On Gizo, people rely on fishing and tourism, scuba diving is big business for some people in Gizo and now scuba divers have lost interest in coming here. But when people from the world came, the aid organizations, we were short of accommodation, every hotel was full so it did benefit hotel owners. (Provincial Secretary Arnold Moveni)

People mentioned the official ceremonies on the anniversary of the natural disaster and church services as ways people remembered those who died and the experience of the tsunami. One church had a listening session to listen to people badly affected by the tsunami and identify ways the church could help. However, most people did not see any reason for further ways of remembering, with some people mentioning that remembering was no use or painful, and others saying that the psychology of the people was to forget it and focus on daily life.

For most people, they thought that the experience of the tsunami meant people were more aware and adaptable to future tsunamis and other natural disasters—such as to move away from the sea and build houses in stronger structures. If ideas were posed that did not seem to take into account the likelihood of disasters then these were challenged:

> Actually government working with NGOs as well came down to us and said to rebuild gardens down by the sea but it might be covered with sea and seriously this advice made no sense. We won't go back to where the tsunami came, in case the time comes when the sea rises again. Every flat area is down near the sea so now the gardens go on the hills. (Religious leader)

However, despite some learning from the tsunami, the attitudes of people were seen as resilient, no matter that the tsunami came and went. People pointed out that the village leadership, ways of making money (apart from a decrease in fishing and increase in gardening), and daily life was largely the same.

Conflict, Peace, Community, and Disaster

People pointed out that the initial insecurity they felt after the tsunami had passed, except for those whose land situation was still insecure. Oxfam was mentioned by local journalists as helping ease conflicts between the Titiana communities by supporting local communities to organize themselves. Otherwise, aid agencies were not seen as responding to the increasing conflict and in fact helping create it.

People thought that while arguments, frustration, and anger about how aid was distributed increased conflict, it had not resulted in an increase in violence in the Western Province. The way people dealt with this conflict differed:

> There were a lot of conflicts, especially when the organizations were giving out supplies. Some people fight. Some people get double share and the people who are more quiet and submissive get nothing. They think just leave it, God will punish them or something like that. It is very peaceful here. There are few disturbances, you can walk the street at night no problems. The tensions were the only time we had worries. Maybe RAMSI is here so it's kind of a deterrent. (Local shop owner)

In the rush to provide assistance, international agencies were seen as actors independent of the government, which then created frustration for people working in local government and further undermined local engagement with the government.

Distribution of assistance to those who asked, rather than based on need, caused some resentment among those who were not aggressive in making demands and changed the balance of power in small communities with permanent effects (e.g., for those who received chainsaws).

Customary governance (involving chiefs and elders) and ways of handling crises were seen as most likely to uphold peace in a time of tension; however, international agencies were seen as working outside these kind of systems and creating alternate power structures, which may threaten *kastom*. In an urban context, churches and local community groups were seen as likely to be the most suitable respondents to crises, and international agencies that provided assistance to these local-level agents were appreciated.

International aid agencies tended to ignore the economic impacts of their presence such as filling hotels and creating accommodation scarcity. Businesses that often suffered losses and became a security net for their employees without shelter felt invisible to international aid agencies and instead used their own networks to respond to the disaster (such as the assistance of overseas family members).

The unresolved conflicts that arose during the post-tsunami period, such as lack of transparency by the MPs on how they spent assistance and land conflict for people in Titiana were seen as long-term problems that aid agencies were reluctant to get involved in.

Western Province remains a nonviolent place, where people feel safe walking around at night, and people know and support one another. In this way, it can be said to be peaceful. However, communities were aware of different meanings for the word peace including perceptions of "peace of mind" and "peace of heart" or a feeling of being "at home" with the community and the community's place in the Solomon Islands.

The initial days after the tsunami, despite the suffering and shock, was said to be days of high levels of cooperation, agency, and in essence peace between community members. Chinese, indigenous, Kiribati peoples helped each other to get food, clothing, and shelter, and there was a feeling of "togetherness" in a crisis. However, the arrival of aid reinforced some divisions rather than built on this cooperation in a peacebuilding sense. In essence, the international agencies were blind to traditional and social support networks in place, and it was only later that some took the positive step of funding local initiatives, such as the church-organized volunteers.

The perception of people was that the tsunami had "opened up" the area to the influence of outsiders—the government and the international community. However, this opening had not translated into peaceful relations with outsiders, with many citing frustration with government, politicians, and aid agencies, and that this experience had people questioning the community's place in the world.

For people whose perception of peace was "peace of mind, peace of heart," or a positive peace, there was less peace in communities because of the ongoing and unsolved nature of the conflicts:

> I am telling you with false expectations, impatience, dissatisfaction about their lives, looking at these promises from government, the members of parliament, then not getting money, things are less peaceful. (Local NGO worker)

Box 4.1   Recommendations for Interventions

Community members in Western Province, Solomon Islands, pointed out that aid had long-term political effects—widening some sources of conflict such as dissatisfaction with the MPs, land tenure uncertainty, and conflict between communities between "haves" and "have nots." Their recommendations related largely to the way aid was distributed in communal societies and included:

Distribute Aid Fairly

- Do not distribute money through the local MPs. This practice was seen to open the way for corruption, favoritism, and misuse of funds meaning that they were less likely to reach communities.
- Aid should be coordinated and based on needs, rather than being given to those who asked, to prevent "doubling up" of supplies in some areas while others missed out. Outside aid agencies should control their competitiveness, which was seen as excluding local actors and unhelpful for communities, local authorities, and local services.
- Interventions should assist with long-term problems leading to poverty and marginalization, notably lack of clarity of land tenure for nonindigenous people.

> **Understand Local Practicalities**
>
> - Start-up money to rebuild homes was seen as more practical than aid agencies rebuilding homes or supply of equipment and materials that may or may not be needed by the community.
> - Local involvement and knowledge is needed in formulating projects so that they are workable and effective.
> - Information and communication was vital so that people could understand what aid agencies were doing and why.

NOTE

1. Oxfam said in a written statement: "It is unfortunate that if as part of the extensive community consultation for this program some people in Titiana were left with the impression that Oxfam had promised them completed houses. This was not the aim of the consultation. Land tenure has been an ongoing issue in Titiana, and Oxfam's policy has been to distribute materials only to those people who have secure land tenure, in order to avoid any implications of encouraging illegal land occupation, and the potential for illegal shelters to be removed. This policy has been explained to the community in Titiana."

REFERENCES

Alasia, S. and Laracy, H. (1989), *Ples Blong Umi: Solomon Islands, the Past Four Thousand Years*, Fiji: Institute of Pacific Studies, University of South Pacific, Fiji Times Ltd.

ANU Enterprise, *People's Survey 2007*, Regional Assistance Mission to the Solomon Islands.

Boege, V., Brown, M. Anne, Clements, Kevin P. and Nolan, Anna (September 2008), "States Emerging from Hybrid Political Orders— Pacific Experiences," *ACPACS Occasional Paper Series, No. 11*, Brisbane: The University of Queensland.

Hameiri, S. (2007), "The Trouble with RAMSI: Reexamining the Roots of Conflict in Solomon Islands," *The Contemporary Pacific*, 19: 409–441.

Moore, C. (2004), *Happy Isles in Crisis: The Historical Causes for a Failing State in Solomon Islands, 1998–2004*, Canberra: Asia Pacific Press at Australian National University.

Naitoro, J. H. (2000), *Solomon Islands Conflict: Demands for Historical Rectification and Restorative Justice*, Update Papers, Asia Pacific School of Economics and Management.

National Disaster Council (NDC) (2007), *Solomon Islands April 2nd 2007 Tsunami: Lessons Learnt Workshop Report*, Honiara: National Disaster Council.

Ride, Anouk (2007), *The Grand Experiment*, Hachette Livre: Australia, Sydney.

United Nations Development Programme (1998, 2006, 2008 editions), *Human Development Report*, New York: UNDP.

World Trade Organization (1998), Secretariat Report, *Summary Observations: Trade Policy Review*. Online. Available at http://www.wto.org/english /tratop_e/tpr_e/tp81_e.htm, accessed on June 19, 2009.

5

KENYA

Sarah Knoll, Vera Roos, Diane Bretherton, and Anouk Ride

Prof. Vera Roos is a clinical and research psychologist at the North-West University's Potchefstroom Campus, South Africa. She has published peer-reviewed papers and presented her research about enabling contexts, relational well-being, and social gerontology nationally and internationally. Vera developed the Mmogo method^{TM}, which is regarded as a visual narrative that assists people to recount experiences and to make sense of the contexts in which they function.

Sarah Knoll grew up in South Africa, and moved to Australia where she obtained her honors in psychology. Her interests are women's empowerment, refugee human rights, poverty alleviation, and peace psychology, with experience in these areas in both Africa and Australia. Sarah is currently a community relations and aboriginal affairs officer for BHP Billiton while undertaking a masters in sustainable mining.

> You see the drought even in the greetings now...when you say "hello," to a Turkana man he says: "I'm hungry."
>
> Pharmacist and relief and development worker Musa Tioko Bwino

> We are going to name this drought immediately after its end. We will give this drought a name, so that in the future we will tell the people what happened. So, people are going to be prepared. So, that they do not suffer the way we did during this long spell of drought.
>
> Chiefs (*wazee*) David Apollo and Moses Allainkori from Katilu

CONTEXT

Kenya is known for its captivating animal species including the big five—lion, leopard, rhino, elephant, and buffalo—as well as a stunning array of plant and birdlife, such as the flamingos on one of the country's many lakes, Lake Nakuru. Its environment is similarly diverse ranging from extreme dry desert to tropical coastal regions, savannah lands such as the Maasai Mara and Samburu to Africa's highest peaks Mount Kilimanjaro, at 5,895 meters and Mount Kenya, at 5,199 meters.

However, Kenya has had a turbulent colonial history, and is now faced with a myriad of contemporary social, economic, and political issues in the country including poverty, political instability, and tribal conflict. These exacerbate environmental concerns posing a threat to long-term stability and sustainability in the region, while water pollution, deforestation, and desertification, caused a loss of biodiversity and hardship for communities. Nobel Peace Prize–winner Wangari Maathai, an international advocate for reforestation and environmental sustainability in the region, has publicized the links between deforestation and poverty, including its impact on rural women who have to walk long distances for food and water.

The Turkana region of Kenya, most affected by drought, is the largest and the least developed district in the country, and lies within the Great East African Rift Valley, covering 77,000 square kilometers (Oba, 1992). Its capital is Lodwar and the district has international borders with Uganda, Sudan, and Ethiopia.

Turkana's mountains, hills, plains, streams, rivers, and valleys tolerate a harsh climate, with low rainfall and high temperatures (91–104 degrees Fahrenheit or 33–40 degrees Celsius). Patchy rainfall distribution creates pockets of vegetation and rangelands crucial to communities' survival.

So, Turkana is not a part of Kenya's largest foreign exchange earnings sector—tourism—which is heavily concentrated around the coastal regions and the large game parks in more southern regions of the country. In any case, outbreaks of political instability, such as the bombing of the US embassy in 1998, and the postelection violence in January 2008, have had devastating effects on the income of local people reliant on the tourist trade.

The economic activity supporting 75 percent of the Kenyan population is agriculture, both for subsistence and for trade. A legacy of colonial times, tea and coffee are the major agricultural exports, but also important are corn, wheat, sisal, and fresh produce (including a recent boom in flowers for export). However, drought, water scarcity, and falling commodity prices have led to difficulties for many of these industries.

The Turkana district lies outside these major economic activities, and most people's way of survival is simpler. They are herders that acquire and manage multiple species of livestock, including cattle, goats, sheep, camels, and donkeys. Sixty percent of the people earn their income from herding or pastoralism, 20 percent from farming and herding, 12 percent from fisheries, and 8 percent from informal or formal employment (*Turkana District Long Rains Assessment Report July*, 2009).

In this environment of water scarcity and drought, the key economic resource for a pastoralist-based culture is mobility. People move between different pastures depending on fodder and water availability, security concerns, and the particular requirements of each species of livestock (McCabe et al., 1985; Little, 1985).

Major outcomes of this lifestyle are food insecurity, hunger, poverty, and conflict. There has also been a recent decrease in condition and market price of livestock, and an increase in food prices and unfavorable terms of trade. Pests and diseases of livestock and crops, poor infrastructure and ability to get food to market, high food prices, raids and cattle rustling, and water-borne diseases further exacerbate the hardship experienced by herders in the area.

As herding (pastoralism) alone is ineffective in meeting the social and economic needs of the local community, people have had to turn to other alternative strategies to supplement livelihoods such as plot farms (*shambas*), irrigation of fertile land, and sale of other items such as aloe, honey, hides, skins (UNDP, 2006) and gum, firewood, charcoal, alcohol (Little and Leslie, 2001), wild fruits, and ethnomedicines. Economic hardship has affected men and women differently (Little and Leslie, 2001), with women and children more likely to explore new income-generating activities for survival, whereas men move to urban centers in search of jobs.

Human development indicators for Kenya also suggest Turkana is not alone in experiencing significant hardship and poverty. While countries such as the Democratic Republic of Congo and Ethiopia rank "low" on the United Nations Development Programme's (UNDP) Human Development Index, Kenya ranks at the "medium" level, coming in at 147, below both Angola and South Africa, but above Sudan. Kenyans have a life expectancy of 54 years, and the mortality rate for children under five of 120 per 1,000 (UNDP, 2009; UNDP, 2007).

In Turkana, infant mortality is higher—159 out of 1,000 children die at birth. Most prevalent diseases include malaria and respiratory tract infection (RTI) accounting for 29.7 percent and 28.7 percent of the top five most prevalent diseases for children under age five in the district (followed by diarrhea, 21.5 percent; pneumonia, 18.2

percent; and eye infections, 1.8 percent). Although the majority of people have not been tested for the HIV virus, it is estimated that the percentage of Turkana people living with HIV/AIDS is 34 percent (District Medical Officer, 2001). According to the *Turkana Long Rains Assessment Report* (2009), the district had recorded an outbreak of polio, with a total of 17 cases reported in three divisions.

Spread of diseases and infections are in part due to the fact that 65 percent of people living in the Turkana region do not have adequate access to drinking water or sanitation, and 28 percent of people are at risk of malnutrition. Prevalence of waterborne diseases, including typhoid and amoebic dysentery are commonly found in areas where communities draw water from hand-dug wells *(akar)* on the riverbed. There are no urban centers in the larger Turkana district with conventional sewerage systems, apart from a few private septic tanks.

In Kenya, public health spending is just 1.8 percent of the gross domestic product (GDP—UNDP, 2007: 24). For Turkana people, health services can be difficult to access—particularly for tribes in the interior. Health centers function around the capital Lodwar, as well as areas along the Kerio and Turkwel rivers, and NGOs such as Red Cross and Merlin have established health centers that provide basic health care. In the Katilu district, Red Cross Kenya coordinates several seminars to educate community members about basic health care.

Similarly, education levels are low in Turkana. Despite the fact, the overall literacy rate in Kenya is 70 percent; in Turkana, however, the government estimates that 70 percent of adults cannot read or write, and illiteracy is considerably higher in females. In spite of free primary education, only 33 percent of 5–10-year-old children actually start school, and the majority—69 percent—drops out before finishing primary school.

Aid in Turkana has generally focused on basic needs rather than services such as health and education. Despite increases during drought relief, Turkana has received less aid than other regions in the country. Prior to the 2005 drought, Kenya was issued around $12.5 per capita aid a year (UNDP, 2004), and currently it is at $34 (UNDP, 2009).

With 38 million people, Kenya's Turkana people live alongside a range of tribal groups, the most prominent being the Kikuyu, which comprise 20 percent of the population; Luyha, 14.38 percent; and Luo, 12.38 percent. Comparatively, the Turkana amounts to 1.52 percent of the population. English and Swahili are the two national languages, providing a lingua franca form of communication between the tribes and other groups. Another commonality in some regions is the significant influence of world religions, with 45 percent of the population reportedly Christian, in addition to sizable numbers of Hindus and Muslims (especially in the eastern and coastal provinces). Such diversity in culture and religion is often seen to enrich Kenya's social environment; however, there are also tensions between groups, and political and ideological conflict.

Colonalization began with the German's protectorate in Zanzibar and then later with the arrival of the British East Africa Company in 1888 and the British colonialism. Farming and plantations followed as European migrants came out from the "Old World" to a new life in Kenya (a lifestyle later depicted in famous films such as *Out of Africa* and *White Mischief*). However, much of this development occurred outside of the northern Turkana region, which, prior to the 1800s, was inhabited by diverse groups of pastoralists, including the Samburu, the Merille (*Dassenech*), and the Rendille. The Turkana entered the region between the second half of the eighteenth century and the middle of the nineteenth century (Oba, 1992).

Despite being newcomers, in the 1900s, the Turkana had huge success in conquering other tribal groups, securing more grazing land for its growing population (Gulliver, 1955). Their fortune was enhanced by the fact that they were isolated from the *rinderpest* (cattle plague) disaster of 1880s and were therefore stronger economically and militarily than their neighbors, whose livestock was depleted. Turkana people continued to trade and prosper at that time (Barber, 1968).

During the nineteenth century, the British were in the process of colonizing East Africa, and King Menelik II of Ethiopia was also attempting to expand his area of influence south and west. The British were also worried about Ethiopian expansion and a Turkana move southwards to challenge white settlers in the Highlands, so they established a local administration headquarters and by the end of 1918, the British had succeeded militarily in pacifying Turkana resistance—often by killing and confiscating livestock. This completely disrupted the pastoral economy, leaving a large portion of the population in poverty.

Just like their pastoral neighbors, cattle rustling (or raiding) is part of the Turkana culture, passed on through war songs and dances. The raids were a means of expanding grazing lands, gaining access to water sources, and an economic strategy of self-stocking cattle and improving social status through acquiring livestock from enemies.

Despite warfare and conflict, perpetual enmity between tribes seldom occurred. Alliances between groups were continuously forged and broken, and marriages occurred between different groups, depending on time and situation. Tribal relationships were dynamic, and members of one tribal group could settle among former enemies who could become new friends, especially during drought and famine. Herein lies an example of the phenomenon of reciprocal assistance, which aided access to grazing and water resources even during difficult times.

However, during the 1920s, conflict began to escalate between the Turkana and other tribal groups. The response of the British was to evacuate the people who lived along the western shores of Lake Turkana. This established a "no-mans' land" and left the Turkana people disarmed and weakened, and yet still vulnerable to unpredictable raids from other tribes.

In the 1950s, as Kenya was in a state of emergency with the Mau Mau uprising and brutal repression of pro-independence groups, the British confiscated all firearms from Turkana. Around the same time, an underground resistance group called the *Ngoroko* started, comprised of retired army personnel and young warriors. Despite intentions to protect and serve the people of Turkana, the activities of this group drove up the incidences of raids and created terror by forcibly depriving people of livestock for rations, and disrupting the economy.

Problems were exacerbated when the British introduced regulations that prohibited Turkana from crossing international borders. This seriously threatened their herding way of life, rendering important pasture and water resources legally inaccessible.

Kenya finally became independent in 1964 under the government of Jomo Kenyatta, amid ongoing conflict with neighboring Somalia. As Kenya moved into an era of postindependence, the new government was reluctant to enforce previous grazing schemes or international border controls, out of unwillingness to be associated with past British and oppressive regulations. Relaxed controls on human and livestock movements increased conflict between opposing groups, leading to an escalation of raids and counterraids in the postcolonial era. Political stability in the neighboring states of Ethiopia, Sudan, and Uganda, and unrestricted border incursions also increased the availability of illegal weapons.

Poverty and disempowerment was further aggravated by environmental degradation as land was overgrazed. Many people stopped herding animals and instead got jobs and traded other goods to survive, as well as resorting to relief camps for food.

Drought differs from other natural disasters that are sudden shocks; it is what is known as a "slow-onset disaster" occurring over a period of time, which is often difficult to precisely define. Technically known as a prolonged period of scanty rainfall in the environmental sciences, drought is to an extent defined by local weather conditions— an amount of rain considered to be a drought in a tropical place may be a wet season in an arid land. Drought, having a somewhat vaguely defined scientific definition, is often socially defined—as an aberration from the way rainfall is "supposed to be" or expected to be. Drought also can be linked to human action (deforestation, introduced animal species, unsustainable irrigation, and farming practices) perhaps more so than other kinds of disasters, and so to a greater degree controllable and predictable by local actors.

In Turkana, water is usually sourced from either Lake Turkana, the Turkwel or Kerio rivers, or offshoot springs from these two rivers. There are also man-made dams, boreholes, and irrigation systems that have been installed through various government and NGO initiatives. The average distance to and from water sources is 4.4 kilometers, and water availability cannot support current demands for human and livestock consumption (Arid Lands Drought Monitoring Bulletin, 2009).

According to the *ARID lands Drought Monitoring Bulletin* (October 2009), the warning stage for the Katilu district (and all zones) in 2009 was at emergency. The Turkana region receives erratic and unpredictable rainfall influencing land use and livelihoods (Soper, 1985). The average rainfall on the plains is 300–400 millimeter on an average year, with less than 150 millimeter in arid areas.

Long rains, usual from April to June, failed in 2007, 2008, and 2009 as did short rains, usual in October to December. This led to a lack of food, long distances to access water, death of livestock, and outbreak of waterborne diseases. Many Turkana people describe the current drought as beginning in 2005 when the entire country of Kenya experienced a famine affecting around 3.5 million people. In 2009, 50 percent of the Turkana people were reliant on food aid (Relief Web, 2010).

RESEARCH OUTLINE

To hear about what it is like living in drought, 14 interviews were conducted in the Katilu, Kitale, and Lodwar communities within the Turkana region of northwest Kenya in November 2009. Interviews enabled the researcher, Sarah Knoll, to explore participants' views, beliefs, meanings, and attitudes regarding drought in real-life contexts.

She first had contact with the community through correspondence with a member of the St. Charles Lwanga Brotherhood, now situated at a Catholic Mission Station in Katilu, Brother Okaye Okedi Francis. This contact was preceded with previous collaboration in the development of projects in urban and rural areas of Kenya, which established a trust relationship over time and through prolonged engagement with communities. A snowball process was utilized to purposefully select people to participate in the research. This meant that interviewees were asked to recommend potential interviewees, who were then contacted and approached to participate in the research, and thereby to ensure that the impact of drought was explored from different perspectives. A brief phone call was made, and a meeting set up for a later time to meet the participants. Due to close proximity of living and working places, the interviewer travelled on foot or by motorbike to meeting places along with an interpreter, to ensure that participant's views were obtained in the contexts in which they lived. The criteria for selecting participants included living within the community during the drought, and having an active and continuing role in formulating and/or enacting the community response to the disaster through involvement in organizing activities as a group or in local institutions such as schools and churches. The selection of the sample was to ensure that the views of a diverse group of people in the community were reflected, such as religious leaders, tribal elders, a chief, a herbalist, community development workers, a farmer, a pastoralist, women, school teachers, and health workers. Interviews were primarily conducted in Swahili (with a translator) and English, and when neither was spoken, a translator was used for the local Turkana language. Interviews were later transcribed into English. Data was analyzed by using thematic content analysis to identify, analyze, and to report themes. Ethical approval for this research was obtained from the participants and the community leaders, and participation was voluntary. The integrity of the findings was ensured by an in-depth understanding of the phenomena by involving researchers, from different subject disciplines, in the analysis of the data as well as to ask participants to verify the findings. The interviewer also reflected on her observations and experiences during the research process by making notes and sharing them with the other researchers.

COMMUNITY RESPONSES

The region of Turkana, in the northwestern corner of Kenya, has always been an extremely dry, inhospitable land, and its inhabitants are known amongst Kenyans for their ability to face adversity, as Daniel Kipsarem Kosgui, headmaster of St. Francis Xavier Atilu Secondary School, Southern Turkana, said:

> There's something I have come to appreciate. The Turkana may look poor, the way they dress, but they have a very strong spirit, you know, they can move on. I always feel if you bought people from the other parts of Kenya here, they cannot survive a week if they were to live the Turkana lifestyle. Sometimes you blame problems here on culture, other times on poverty, but I think it is poverty. Even here in this community, poverty is really biting, along with environmental degradation.

Turkana people see drought as a prolonged period of time where the weather is extremely hot, where there are few clouds in the sky, and when rains fail to come. Drought is regarded as an ongoing challenge because it unfolds over a period of time rather than being a sudden disaster:

> They are saying that the drought started early this year. Simply because the short rains failed last year. They are talking about round about 3 years. (David and Moses, two village elders)

> Previous years it would take six months without rain, and then the seventh month, there would be rain, but since this year, three years, there have been no rains. (Awoi Akuwai Elikweli, *emuron* (herbalist/rainmaker))

Drought was predicted by Turkana people by observing changes in nature as well as by applying indigenous knowledge:

> The signs, there is no grass, the water points become dry, when the animals themselves give birth, there is no milk, they cannot feed the babies. Also, the old *wazees* [chiefs], the old men, they are able to predict, by observing, the drought is coming. (Chief Alan Lokeyun, chief of the district of Katilu, South Turkana)

> People were reading the skies, looking at the skies....Well, people, look at the stars. There are some stars, when you look at them, if it reaches, probably April, and there is nothing, that is when it should start, the rainy season. So, people start worrying, now things are going to be bad. People can see the stars are not moving in the same way, things are not moving the way we expected. (Chris Ekuwom, project manager at Oxfam in Lodwar)

> They are saying, those men slaughter a goat, and now, the intestines, they don't know how they do it, they can predict from the intestines. (David and Moses, two village elders)

The effect of the drought is most acutely described by the people in the way it impacts on the environment around them, wreaking widespread devastation of plants, animals, and people:

> What did I think of [about the drought]? . . . Death. I thought of death. It is something that has happened. Once a person does not get the food, then there is death. And also the diseases, like *kochioko* [kwashiorkor, a form of childhood malnutrition] for the small children. (Elizabeth Emunyen, teacher at Turkana Integrated School for the Blind)

Drought has been a constant, chronic feature of existence, and people cannot remember a time when drought did not exist. Many have reached a level of resignation, incorporating it into a life characterized by hard work, suffering, hunger, and constant searching for food as they move in quest of pastures and water. Drought was a part of daily habitual conversation. When asked what people were doing when the drought began, the concept of time is almost meaningless:

> I was here, and I was doing nothing. I had nothing to worry about, because if the drought is there, there is nothing I can do. I just wait for the drought years to go. If it does not end, I will just do nothing. (Awoi, *emuron* (herbalist/rainmaker))

Traditionally, the Turkana are nomadic, and so the constant search for pasture, water, and habitable living environments has always been a way of life, as has their reliance on cattle and livestock for food, capital, and social status. Their interrelations with other tribes, experience of conflict and peaceful coexistence, and trading practices have also been inextricably linked with changes in weather and environmental conditions.

People's experience of drought extended back to before they were born, in the 1960s, a time when international relief operations were launched in response to the extreme drought conditions in the region. Experience of the drought was passed on through stories from generation to generation, and the stories were symbolized by special names given to each drought:

> That one was the very worst, in 1960. *Namotor* [the name given to the drought] actually is what brought the agro-pastoralists to where they are, now. Although I wasn't born at that time, 1960, I am told that the people now, who are living in Katilu, came there from Turkana north. The others now, they are in Turkana south, also came from the north. Others came from the western areas. At that time, that time of severe

drought, the locals they went to the shores of Lake Turkana to survive. (Chris, development worker and pastoralist)

There was a bad drought in 1980. They call it in Turkana *Lochoo*. A lot of people died, people have made a lot of songs about them. When they come together, they sing those songs at a ceremony, then they share some food, and they remember. What happened, in *Lochoo*, all the wild animals, the buffalos, the elephants were found around here looking for food, for water. Most of them died, and also the people also had nothing to eat. So, they remember these things in the songs. (Elizabeth, teacher at Turkana Integrated School for the Blind)

Telling stories and singing songs brought people together to share the hardships people endured in the past. Also, it was a means of expressing the solidarity of the Turkana people and their culture, of how they had faced adversity. Lessons from the past could then provide motivation to look to the future together as a culture and community:

They sing songs about hunger and it can help them, because it reminds them this thing [the drought], it goes, and then it comes back. (Pastor Sammy Meli, local Pastor for Katilu district)

Despite increased access to external education, knowledge, and resources, the Turkana way of life is still heavily based on custom and cultural practices. *Emuron,* the traditional medicine men or rainmakers, are highly regarded by the community, and have significant influence on their beliefs and practices. *Emurons* are believed to be responsible for predicting weather patterns, as well as controlling environmental conditions and the fate of the community:

The *emuron* can tighten the knots to make it so severe, so that is how they must now respect them....They say the *emuron* is a middleman between the gods and the people. (Lopeyok, "mamma" or local woman active in community activities)

But the rainmakers, the witchcrafts, they made people come together, on the belief that they can do some rituals for rain to come, but it seems the rain did not appreciate what they did, it did not come. (Elim Shadrack Lotonia, Red Cross medic)

People engage in dances, prayers, and rituals in an attempt to reverse this misfortune:

The community calls the village elders, then they request an old man, who is an *emuron*, and then they have special prayers. Thinking that God is going to hear them, and send the rain. You put water in a calabash, and you go, sprinkling water over the animals, during the celebration. (David Apollo and Moses Allainkori from Katilu)

You know sometimes the old men they come together, under one tree, and you know, there is this one prayer that we say in Turkana, *Agata*, they get a very big goat, or ram of three years old, then they call upon all sort of names, some curses, they curse those things, any bad omens there, then they call upon God to bring the rain. Then they roast the ram. Then one may pray, cursing on the bad things, calling upon the blessings, and what have you. (Elizabeth, teacher at Turkana Integrated School for the Blind)

Community responses to the drought are often based on their memories of what had happened in the past. Negative thoughts and emotions such as fear and anger are evoked, as people are anxious about their livestock and the livelihoods of their families and community:

Yes, of course [I was scared]! Because, even your own children...you know you would be scared, if there is no food, if there is no water. (Lily Rose, shopkeeper)

When you hear about drought you know it is coming to sweep out your livelihood. They are being threatened; you can get very angry with that kind of phenomena which is coming. (Chris, development worker and pastoralist)

The signs are starvation, and starvation—more, and more animals die! [We] feel angry. Because people are dying, animals are dying... (Milton Loito, farmer in South Turkana)

Sometimes people looked for others to blame—the neighboring Pokot tribe, or blaming the local rainmaker or shaman, or others in the community for the disaster. Daniel, headmaster, described a typical case:

Yes, in fact, I heard one time one of my workers, when I asked him why it is so dry, he said that it is rumored that one of the rainmakers set up a snare to catch an antelope, and someone took the snare down, and so as punishment, the rainmaker chased the rain away.

Sometimes blame for this event was attributed to a mix of sources, but most people agreed humans were somehow responsible:

OK, I do not blame God totally. But I do blame him, simply, because the situation is very severe. So, I just need God to do something. I hear in some other places, it is raining, so why is this place just dry? Also,

I blame my fellow men, because they are cutting down the trees. Ya, because they are burning charcoals. (Chief Alan, chief of the district of Katilu, South Turkana)

There was little variation in people's responses as to who was most seriously affected by the drought. Older people were identified due to inability to move, find food or water, and their dependency on other family members. Immobility and dependency were also offered as an explanation for why babies and young children were vulnerable to negative effects of the drought, as it interfered with adequate development and health. Financial hardship also affected schoolchildren because parents were less able to pay school fees, interrupting education or reducing their chances for education. Women were also affected heavily as providers of food for the family, including older persons and children.

Despite the hardship, people assisted each other to get food and basic necessities. Community and solidarity are meaningful concepts in Turkana culture, fostered through a collective response to adversities, such as sharing:

> This is something I realize with this community, when someone comes to your place, you cannot deny, then you have no alternative but to share. (Daniel, headmaster)

> When a person has died, they go there together, they share together, they nurse one another, and all those things. OK, and also at times of drought, you get something, you harvest something, there is this idea of sharing, although you are not given some big thing, you still...share. (Elizabeth, teacher at Turkana Integrated School for the Blind)

> Drought is a time of stress, so I could go to town, I get food from relatives and then maybe I go and assist another person, we all support each other....I will support friends, of course, because I fear that in the future I also might need assistance. (Chris, development worker and pastoralist)

A key form of communal decision making to respond to droughts was the *barasas,* traditional meetings of different stakeholders including chiefs, tribal elders, and community members. These are used as time to identify issues, communicate opinions and alternative solutions, disperse information, and facilitate mobilization strategies. *Barasas* are also utilized by the NGOs and the government when requiring the cooperation of the community.

> Immediately I realized what the situation is, I called the people for a meeting, the villagers, and told them: "Understand that, we are not going to harvest anything, because the rains have failed. So what

you are going to do, please go and ask your neighbors, who have the *shambas* [small farming plots] along the river, to go and halve a section to cultivate. You can come and fetch the water from the river, and you can come and pour it on the crops.... You don't only have to rely on the farming, and the livestock. You have to take the children to school." If the people are educated and capable they can start the businesses and farming. (Alan, chief of the district of Katilu)

People in higher-status positions play an important role in facilitating an organized response by advising people on what to do, and supporting them in their endeavors.

For me, I tried to advise my people, if there is any way of surviving, do it, maybe draw some water from the main river, to whatever they can do... and whatever they have, to keep it, not sell it. (Pastor Meli, local Pastor for Katilu district)

Farming has been a prominent response of the community to the drought, and it has enabled the community to adapt and survive in the face of widespread devastation. When asked what their initial response was to the drought, a man and his wife who tend a *shamba* along the Turkwel river responded:

When it started, I called those people who use irrigation, I told them let us go to the farm, and start farming...The action I was taking, I was telling everybody to come to the *shambas* and do the work. (Milton and wife, farmers)

Local churches, schools, and other organizations promoted diversification as a response to the drought:

I would like to train some groups, so that we know they are a source of income for their families, I would like to train them to practice a diverse way of living. So if they have a *shamba*, then they also have some livestock, and also some small business, or even having something else somewhere they can help them. And also, telling them to take their children to school, and also when their children can be employed, that is another source. (Pastor Sammy)

But now you see things are changing. But instead of every time bringing food, they are now doing other things for the people to depend upon, especially like agriculture, about the livestock, and the education to change everything. Once you are educated, you have ways of changing life. For the better. Ya. (Elizabeth, teacher)

We are trying to keep poultry, milk, goats in the school, to make them see that they can reap the benefits. We try to make them see that agriculture is all about getting your hands dirty....We also began an environmental club, where we encourage the youth, not only in the school, but also in the community, to plant trees. They were responsive. Very positive. Right now, we have been involved with the government with a group called *"Kazi kwa vijana"* [work for youth]. I remember we were talking with the district forest officer to use these boys to plant trees, at all the schools, the dispensary, the hospitals, to plant and then to look after them for a little while. The government funded it, and it was led by the community. The provincial administration, then the village elders who fall under the chief, they help to administer it. And very much the *mammas* [mothers]. (Daniel, headmaster)

When people found alternative ways to survive, such as agriculture, selling of charcoal, wild fruits, and alcohol, as well as starting businesses, families and communities had to stay in one place for a while and change their usual herding activities. Not everyone was open to change—moving from pastoralism to agriculture, was a major lifestyle shift:

According to me, maybe they are afraid, it is difficult to adapt to this kind of life. For example in the towns, maybe someone is looking for work, waking early in the morning, looking to the *shamba*, to plough, early in the morning. They are not used to this life. They are used to a life where they sleep, maybe till, eh, 8 o clock in the morning, then they take their animals, they come in the evening, life without any obstacles, no one directing them, and they don't want to be directed. (Elim, medic)

Much responsibility is afforded to the *mammas* (women) of the community in terms of seeking alternative income-generating strategies, labor, and mobilization:

I started talking to them. The river was available, there was water. We talked of the little small *shambas* along the river, to grow some food. And also trees, which were dried, they could burn to charcoal to sell. So, because I was working, I took the little that I could get, to share with the others who are suffering. Something little that I sold to people, I sold them at a low price to people, just so that they can keep on going. Some of the women, they went for wild fruits. You know the trees, which can survive in dry areas, they collect the wild fruits and sell it. Also when she goes some distance looking for some wild fruits, like *edapal*, they boil and boil, and when it gets a bit OK, they eat. [She works] very, very hard! Women are like the termites. Eh, working here and there, all the time. There is no day when you will see a woman not working! (Elizabeth, teacher at Turkana Integrated School for the Blind)

They begin, a group of women, they started this week, a bit for me, then next week, they got a bit for this one, the other week...so that we keep on assisting one another....Some mammas here, like now, some they sell their tobacco, this small tobacco. They start businesses, so that they keep putting the small profits they are getting, so that they put together with the other ones, it can also help her, to buy a goat. (Lily Rose, shopkeeper)

Women said that not only was the responsibility physically demanding, but that there was also a stronger, more powerful emotional burden, the suffering of children and the community. Still, the drought also promoted action, solidarity, and opportunity:

[The woman] works so hard, to fetch the firewood, to sell for thirty shillings, and that one is not enough to feed the family, the children, you know. The children do suffer. The woman is so affected because she sees her children suffering. She can do nothing. You know, the women do things together, for instance, when they go to fetch water, they do it together, they talk about the family issues, they talk about the burning issues, they release....Nowadays the women have gotten posts in the community, and also in the places of work, women have come up, they can hold the good posts, and they are the ones who are making things go peacefully. (Elizabeth, teacher at Turkana Integrated School for the Blind)

The drought poses a challenge to the cultural norms and assumptions of the Turkana tribe. To some, these changes are viewed as positive. To others, they are viewed as degradation of culture. For many the embeddedness of certain ways of life such as keeping and amassing cattle as wealth made it difficult for people to be aware of their role in perpetuating drought and to adjust their behavior. People of Turkana were battling with the question of what extent to change to survive and limit risks of adversity whilst maintaining their cultural identity:

Livestock is in the centre of their lives. Without livestock, there is no life. And events of your culture revolve around the livestock, like even the marrying, doing initiation, going to raid, cattle rustling to do restocking of livestock, they come, they kill you for the livestock. It's a complicated way of life....Normally, those ones who are stricken by drought, they come to towns, take children to school, completely change their lives. Some become doctors, engineers, and then their parents are in town. You look at them and think "they are not true Turkana."...One very funny thing about pastoralists though is even though they go and get jobs, and completely live a so-called "western life," you find somebody in the family must keep animals. There is a spirit within them that says, "No, I must keep animals, I must keep livestock." (Chris, development worker and pastoralist)

They believe in animals, and if you don't keep animals then you are nothing. And the drought is affecting the keeping of these animals....And the problem is also with eating vegetables. You want them to eat vegetables, but they only believe in eating meat. So it is affecting the life of the people. (Brother Simplicius Wandera, Katilu, South Turkana)

In this culture, there is an almost worship of wealth. And their wealth is in terms of animals, not money. The elders graduated through the same process, it is almost as if the youth are emulating them. (Daniel, headmaster)

It is a good thing to cultivate the land. But it is good to have both animals and land. It assists for them to have in case if there is no food in the land, they can have food from the animals. (Awoi, *emuron* (herbalist/rainmaker))

Cut off as it was from major markets, infrastructure, economic enterprises, and government services, Turkana's drought highlighted the many unmet needs of community development. One resource that communities needed but did not have is adequate education and health services said Elim, medic:

There have been a lot of cholera cases in this district, I have been moving around the district, 77,000 km squared, and I have not even achieved all of it, I get calls, people are dying of cholera, and this cholera is being caused by not just contaminated water, but also animals dying because of drought, then the community, because of their lifestyles, they eat the contaminated carcasses of these animals, which have stayed there for 4 or 5 days....We need enough medical professionals to cope with these diseases that come. Drugs, personnel, resources. To prepare for the drought when it comes.

For Pastor Reverend Meli part of the problem of human suffering in the drought was lack of agricultural development:

So, in drought, what I think the community needs, they need maybe to have, to build a store, so they can store excess things. They also need to have a livestock market. They just sell their stock where they are, for very low prices. But if there is a market, they can sell at a good price. And then another thing, maybe they need farming implements, instruments.

Information, irrigation facilities and techniques, seeds and agricultural implements were identified as things people required to farm but did not have. There were no plans, based on environmental knowledge, to deal with drought and land usage in South Turkana, Chris, development worker and pastoralist, pointed out:

> People should actually terminally move out of their places where they can access pasture and save animals. Then the rangeland management should plan: "Ok, let's graze our animals here, leave this one, then in a certain period of time, we move to this range land when that one is depleted." So, they can try to cope with the drought but it has not happened. Because right now someone wakes up and says: "I will take my animals there," and who can control them? They are moving to Uganda, and you know when they are met with resistance, they will fight. The pastoralists say: "We don't believe in these international humanitarian borders!"

For many years, external aid in the form of government and NGO assistance was predominantly food aid. At first, people were happy to receive food in a climate of starvation and suffering. However, coordination efforts were inconsistent, and communities came to realize that food drops were unpredictable. The view became that NGOs were unreliable:

> NGOs have always been temporary people. They come for a day or a week, and then it is all forgotten. (Ereng, "mamma" or local woman active in community activities)

> NGOs come, you give me today, tomorrow you are not there. You have not helped me. And most of the NGOs, that is what they are doing. (Brother Simplicius Wandera, Katilu, South Turkana)

There was a sense that aid was being supplied without a view toward solving the problem of poverty and hardship:

> They brought relief food. They are talking about around February, this when the community came in. The food itself is not continuous, sometimes it is there, sometimes it is not there. So [the community] are not comfortable with it, they cannot rely on it. It is there for six months, then it is gone. (Milton, farmer)

You know. Food cannot help at all. Unless they change food to something else, you know, unless you tell someone, "this is how to make food," that is so important. This person will go out and do it, instead of thinking "tomorrow, food is coming" and so they do nothing. (Elizabeth, teacher at Turkana Integrated School for the Blind)

We do not have so much belief in relief food. It will only come and solve the problem for a few days, and after that, you still go back to the same situation. We need to get new ways of doing things—get more *shambas*, or more land to cultivate. Because it seems, from the farms, even if the drought is there, there is something we can eat. (David and Moses, *wazee* (tribal elders))

By the way, the people are not really happy about the relief food. They know. They know that thing is unsustainable. And, again, currently, they have actually changed their perception. You get the women groups, the youth at risk, they feel that, can you give us money, so that we can do something we feel is very important for our lives? Rather than just giving us food. Another saying is, can you change this food into money? Because we can do something. You know, there are some development workers, who believe that relief food is the best assistance we can provide. It is not. Relief food should really be brought in when other food production systems have collapsed. But immediately, once these production systems are actually working, then it should be phased out. We have to link relief and development. Because whatever you do in relief time, you should actually build into development work. And that link should really be fortified. (Chris, development worker and pastoralist)

Ya. They don't ask the people what they need. Another problem is, they put much stress on trainings, instead of giving them a start to get an income. I would like to train some groups, so that we know they are a source of income for their families and they have diverse ways of making a living. So, if they have a *shamba*, then they also have some livestock, and also some small business, or even having something else somewhere that can help them. (Pastor Meli)

Various NGOs and government agencies have invested in building water-storage facilities, providing agricultural education and training, as well as resources in the form of seeds and farming implements. NGO assistance included food aid, income-generating schemes (microfinance), irrigation schemes, boreholes, water tanks and water pumps, "Food for Work" programs to plant trees, build dams and dig trenches, farming assistance, education, and destocking programs. Oxfam and other organizations (including VSF or Vétérinaires Sans Frontières, Belgium) began a program known as "destocking" or buying animals from the Turkana in an attempt to assist the economic hardship of people and ease the environmental burden of so many cattle. This had some negative effects as markets were distorted, and people used the program to fuel conflict and competition, as one community worker explained:

> They [the NGOs] have reduced the over-stocking of animals, buy-ing cows from the community, and then donating them to hospitals, to slaughter for the community for food. There is still the issue they sell the ones they feel cannot survive the drought. I think it [selling animals] has increased, because when they (the Turkana) came the first time, they sold their stocks, and this forced them to go to the other side (the Pokot people's land) and steal more for when they come again.

Where solutions that emphasize sustainability and empowerment have been promoted and effectively implemented, the community responded positively recognizing that this kind of preparation, edu-cation, and provision of resources lead to long-term solutions to the problems they face:

> Yes, they responded well. They saw it actually sustained some commu-nities. For example on the other side where Oxfam is, the communities actually felt a small impact of the drought, because Oxfam had pre-pared them, given them a lot of resources, in terms of water, boreholes, handpumps, for their livestock. (Elim, medic)

> I think that when they start projects, they should be sustainable, and then they should be included in the management, be made more inclusive. Because I saw one time, with the irrigation scheme, they were supposed to plant, the chairman dictates, but everyone must wait for the chairperson who has not opened his farm. And then the others they see nothing happening so they give up and go back to their animals. (Daniel, headmaster)

Other positive impacts of external aid include increased access to education and employment opportunities for the community dur-ing drought. Many parents, having lost their means of survival when livestock died, were forced to send their children to school, which had feeding programs. As a result, many community workers, such as doctors, teachers, and agriculturalists, have returned to their com-munities equipped with education from larger towns and cities.

> So, this drought finished all of our animals, so we couldn't herd a sin-gle animal, so in one way, it assisted my parents to take me to school. Well now, because they had nothing to depend on, and they had a

feeding program at the school, so we went to school....Yes, I think that is how most children go to school. It is a good thing because this is the only way that can change the community. A family that has taken some children to school, they will not be affected so much by the drought, because even if the animals are affected, people can find income and will be taken care of. (Elim, medic)

Still, Elim highlighted ongoing resistance to education and the effects of illiteracy:

This is a community who has got its education late in life, so most of the community is semi-literate. So it is difficult to convince them, because if we had enough people who had gone to school, they would have changed the community, given elder education, civic education on how to survive, on how to cope with calamity, but now because of the portion, not even a handful who have gone to school, most are illiterate.

Education is seen as an instrument in bringing people out of poverty, of generating change from within:

Actually, I have also come to realize that you cannot bring an outsider to change them. It should be through education, and in the long run it will change....Before, the elders [chose] the strong boys to look after the animals, they call it "*wote*," and the weak boys who always got lost, they were the ones who were sent to school because they were [considered] useless. And eventually it turns out to be the [boy] who was neglected, who comes back from school, they can secure employment, and realize they can be better than their older brothers, they build nice houses. They are starting to realize that cows are not sustainable. (Daniel, headmaster)

Inevitably, education brings the generation of new ideas, exposure to concepts, ideas, and cultures outside of one's own posing a challenge to balance development and survival with traditional ways of life.

Education has also had a significant impact on women. Not only are they able to perform their responsibilities to a more proficient level, but also they are able to utilize their hard work ethic and become instigators of change and positive role models within the community.

Here, [it was believed], the girls, they are a waste of time to send to school. And also, they do not make any decisions. They can just be married off to any man who has the wealth. Hopefully now, if they can be compelled to take the girl to school, they will see the difference. We can see some of the ladies who go to school go to college, and one is even aspiring to be a member of parliament. (Daniel, headmaster)

Ya, I think that if you check our relief registers, you find that it's the women who are actually the heads of the households. The reason being, the women are the ones who can really manage the food, the good managers. (Chris, development worker and pastoralist)

"Relief food" and "NGOs" have become common words in the Turkana language, and most respondents have grown up having been impacted in some way by this development. Increasingly, people believe that assistance can only be effective if the community is involved in the process, and there is a push for increased community consultation and engagement in development project design, implementation, and evaluation:

Of the NGOs and government people who come to help, we only accept those who will come to give them skills, or give them assistance on how to get food, on how to help themselves. These people providing food, they never come to discuss the community's priorities. So they just come and say: "What do you need? Maize? This is maize. Have it!" I would appreciate it if people would come and ask us what we need. And then I would tell them, actually we need somebody to assist us to irrigate their land. Second, people need somebody to restock the animals that have died. Enemies have come, and they have taken our animals. But all these things, well, we have never been given a chance to express ourselves. (Awoi, *emuron* (herbalist/rainmaker))

We were very grateful, we were happy that someone came in when they were in need...but still dialogue is not here. We get what they think the locals need—that is what they give in reality. The people, they could have other priorities but the aid workers, they don't care. (Alan, chief of the district of Katilu)

No, the NGOs didn't consult the people. Somebody in an office thought it would be a good idea. One, it is not sustainable, and two, you are reducing people to waiting, for somebody to come, instead of showing them how to do it themselves....I think one major weakness, the NGOs have been helping, but they have reduced people to waiting for them to come up with solutions. I think that when they start projects, they should be sustainable, and then they (the community) should be included in the management, more inclusive. (Daniel, headmaster)

Some of the NGOs, they usually listen to what the locals are saying, but some of them, they don't care, they have to do what they think is the best. (Milton, farmer in South Turkana)

Governance and transparency issues exist at the local, ministerial, and government levels. Relief food and money is susceptible to corruption and embezzlement, and resources are not distributed in an appropriate and accountable manner. This is perceived to contribute to mistrust and conflict between local-level administrators, provincial ministers, and government officials, and obstruct proficient implementation and long-term sustainability:

> I think they do get, but not the quantities which were originally allocated.... It is pilfered on the way. Starts with the big man in charge, who gets his share. (Local community worker)

> You know, once you are corrupt, nothing is going to happen to you. (Local service worker)

> The obstacle here is just the local leadership. Because these are the people who are used to giving themselves. So when they see he is coming with this approach (empowerment, sustainability), they feel bad. They start telling people, this is not good, it shouldn't be that way. [The] government, councilors. And the chief elders. These are the obstacles. Because these are the people who have stayed in the government for over 20 years. So they don't like this approach. (Local health worker)

> I attended a meeting with one international NGO recently, I think some of their decisions and expenditure are questionable. They use US dollars, it is difficult to understand. They have community representatives who are supposed to be on the board. But rarely do they come to the ground and speak to the local person. To see what they really need. The government, for example, brought a tractor, plough and trailer to help with the irrigation scheme. But the same old problem with mismanagement came up. Now, the tractor works, the money comes, and a small group spends the money. Then when the tractor needs new tires, the money is not there. (Local religious leader)

Community-valued programs that promote self-sufficiency, sustainability, and empowerment and threatened the capacity for corruption were seen as a way forward for development and assistance.

CONFLICT, PEACE, COMMUNITY, AND DISASTER

Historically, Turkana has been a "closed district"—shut off from resources and adequate infrastructure, development, and prone to cattle rustling and poor weather. Changes caused by drought and assistance has opened up traditional settings to new people and new ideas, but customary ways of using the land even if environmentally harmful, were difficult to change.

Economically, some outsiders took advantage of the drought, which created resentment:

> Well, for example, this is a non-agriculture area, so, we depend on other communities to produce food, and then we buy. Then we have our animals, livestock, so livestock you cannot eat them all year long, and we don't have market for our livestock, so we are limited ourselves. So, instead, you now get some middle men to buy at a cheaper price, and they take it to meat industries, and sell it at a good price. So now, we go and sell a goat at 500 shillings, and he goes and sells the same goat for 2,000 shillings, so now there is exploitation. (Elim, medic)

Inside communities, stealing, corruption, and disputes over relief food, whether driven by desperation or opportunism, led to conflicts:

> *Hakuna amani* [there is no peace]…there is no peace. First thing, during the relief food, there are those people, because the relief food is not enough, there are those who are going to benefit, and there are those who are not going to benefit. Then there are those with the *shambas*, the gardens, and they have already cultivated, who go there to steal what is already there. Others, who are grazing, take their livestock inside there (to the *shambas*). So insecurity, it's very high. (Chief Lokeyun)

People complain that often relief food falls into the wrong hands, and those who really need it do not receive it. Community leaders are responsible for curbing the conflict.

> The relief food becomes acrimonious. There is always the fight that this person should not get the food because they are well off already. These disputes rarely go beyond the tribal elders. They are reprimanded, and then it is resolved. (Daniel, headmaster)
>
> Sometimes, when it started, you find administration, they start to gather, start talking, people have suffered…people get some food who are not unable, but those who are hard working, they don't get enough food to sustain themselves. (Lily Rose, shopkeeper)

Sometimes, it was difficult to discern who needed the assistance, with some people making dishonest claims to relative deprivation:

> But now you find, even the people you find, for instance, some people here who have large herd of cattle, camels, when you greet them, they will give you the same response. Then you find that old lady, who is

here [indicates backyard of compound] if you greet her, you will get the same response. So, to find out, if you are not a local, to gauge between who is most vulnerable, or in need, you cannot do it. (Musa, development worker)

In traditional Turkana culture, the nuclear family is called an *awi*, and each *awi* manages its method of survival. Sometimes a group of families might cohabitate under one leader, and this is called an *adakar* (Akabwai, 1992). The formation of an *adakar* allows for negotiation of water and pasture rights with neighboring tribes, and increases security when moving to neighboring territories. While some people identified *adakars* that had become more adaptable and settled on land by the rivers relatively peacefully, for Turkana generally it was seen that the drought had exacerbated violent conflicts of *adakars* and the Pokot tribe.

Drought encouraged people to move in and beyond Turkana borders as scarcity of natural resources like water and pastures increased. Young herders were armed and often fought over cattle and grazing rights. As a result, most people associated the drought with more conflict:

The past three months, we have seen an increase in conflict with the neighboring community. Especially raids. When people lose their stocks, they must replenish, they must restock. The only way they can do that without paying anything, is to raid. The rainmakers dictate when they go for raids. They are the same people who bless the warriors. It is almost a ritual, you cannot just wake up and go and raid, you must be told by the rainmakers the conditions are right. These rainmakers, or *emuron*, they said the rains will come in three months, and suddenly the rains did not come, so there is more conflict...and interference in development, even with the irrigation scheme, men come in and break it. It is difficult to make progress. (Daniel, headmaster)

The community here are fighting, because the Pokots come and steal the animals, the Turkana go and bring them back, so that is also another problem. It's the Pokot. Because you know they say, "Take them from us, feed them for us, and then we will come and steal them again" [laughs] so that is what is happening. (Brother Wandera)

HAKUNA!...There is no peace! [laughs]. There is no peace between them and their enemies. Ya. Even the government is trying, but there are no payoffs. If any government will come, they will need to close all the borders between Turkana and Pokot with the army forces. And then, the other thing is, they need to do disarmament. Which is difficult—if you disarm people more guns will come, you know from across the borders, Sudan, Uganda. (Awoi, *emuron* (herbalist/rainmaker))

Last year, Turkana people moved to the Ugandan National Park, and they were so hungry, they started killing the wildlife, and eating it. The Ugandans didn't comprehend it, they didn't understand it is because of the drought, so then they killed them, bombed them, so actually they became a victim of conflict. (Chris, development worker and pastoralist)

Raids had been a traditional way of amassing prestige and wealth, but the consensus was that the raids became more severe at the time of the drought. Peace initiatives such as bringing leaders together, establishing an agreement on natural resource use, education, settlement, and agricultural development that cut competition to amass cattle were seen by the community as important to solve this problem, as was border security.

Box 5.1   Recommendations for Interventions

Drought is a different kind of natural disaster than those that are sudden and unexpected. Therefore, recommendations from Turkana people focused more on the need for intervenors that could help adaptation to the disaster, rather than band-aid solutions to the problems they faced, such as the following.

Support Local Adaptation

- It is important for outsiders supporting a community in hardship to discover the strengths of the community.
- Change should come from community's strengths—communities have their own coping strategies, which should be valued and respected by outsiders, and helping organizations. Organizations suggesting changes, to adapt to the challenges of the environment, should realize that the people and their cultural norms and assumptions cannot be changed from outside the community. Communities regulate their actions and processes to protect its members and its own survival.
- Helping individuals and organizations should use the existing communication systems in communities to promote the community's adjustments to the disaster. (There is a recognizable hierarchical structure to disseminate information through village elders and *barasas* in Turkana.)

- Education can enable employment and transference of technical skills to help people cope such as farming methods so people could diversify family income.
- Health-care staff in the region is needed to prepare for the disaster and the impacts on people's health.
- Diversification of income initiatives need to be sensitive to the cultural changes required by Turkana people to adapt to new livelihoods.
- Disaster plans and land-use planning are important to prepare Turkana people for future droughts and encourage adaptive and preventative strategies.
- People want relief and development activities to complement each other to provide Turkana communities with sustainable strategies to cope with drought and make a living over the long term.

Support Peace Initiatives to
Increase Community Capacity

- Peace initiatives that enabled the Turkana and the Pokots to come together and settle disputes are desirable, along with measures to limit the spillover of conflict and weapons from other states into the Turkana region.

REFERENCES

Akabwai, D. (1992), *Extension and Livestock Development: Experience from Among the Turkana Pastoralists of Kenya*, Pastoral Development Network Paper 33b, London: Overseas Development Institute.

Barber, J. (1968), *Imperial Frontier: A Study of Relations between the British and the Pastoral Tribes of North East Uganda*, Nairobi: East African Publishing House.

Braun, V. and Clarke, V. (2006), "Using Thematic Analysis in Psychology," *Qualitative Research in Psychology*, Vol. 3: 77–101.

Ellingson, L. L. (2009), *Engaging in Crystallization in Qualitative Research: An Introduction*, London: Sage Publications.

Gulliver, P. H. (1955), *The Family Herds: A Study of Two Pastoral Tribes in East Africa, the Jie and Turkana*, London: Routledge and Kegan Paul.

Lincoln, Y. S. and Guba, E. G. (1985), *Naturalistic Inquiry*, Newbury Park, CA: Sage Publications, Inc.

Little, M. A. (1985), "Multidisciplinary and Ecological Studies of Nomadic Turkana Pastoralists," Biology International (IUBS, Paris), 11: 11–16.

Little, Michael A. and Leslie, Paul W. (eds) (1999), Turkana Herders of the Dry Savana, Ecology and Biobehavioral Response of Nomads to an Uncertain Environment, USA: Oxford University Press.

McCabe, J. T., Dyson-Hudson, R., Leslie, P. W., Fry, P. H., and Wienpahl, J. (1985), "Movement and Migration as Pastoral Response to Limited and Unpredictable Resources," in E. E. Whitehead, C. F. Hutchinson, B. N. Timmermann, and R. G. Varady (eds.), *Arid Lands: Today and Tomorrow*, Boulder, Colorado: West View Press.

Murekefu, W., Mbiuki, A., and Mbolu, K. (2009). *Turkana District Long Rains Assessment Report: Regional Assessment Team of the Kenya Food Security Steering Group.*

Nieuwenhuis, J. (2007), "Qualitative Research Design and Data Gathering Techniques," in K. Maree (ed.), *First Steps in Research*, Pretoria: Hatfield.

Oba, G. (1992), *Ecological Factors in Land Use Conflicts, Land Administration and Food Insecurity in Turkana, Kenya*, ODI Pastoral Development Network Paper 33a, London: Overseas Development Institute.

Office of the Prime Minister, Ministry of State for the Development of Northern Keyna and other Arid Lands (2009), *Drought Monitoring Bulletin* (Reource Management Project II) Turkana District. Online. Available at http://artikelpdf.co.cc/link/drought-monitoring-bulletin-july-2009/, accessed December 2009.

Relief Web (2010). Online. Available at http://www.reliefweb.int/rw/rwb.nsf/db900sid/MUMA7V38HS?OpenDocument&RSS20&RSS20=FS.

Soper, R. C. (1985), Socio-Cultural Profile of Turkana District, Kenya: Institute of African Studies, the University of Nairobi.

Turkana District Long Rains Assessment Report July 13th–17th (2009), Kenya, Wycliffe Murekefu Ashford Mbiuki Kithama Mbolu, Turkana DSG Members Ministry of Livestock Development, World Food Programme Ministry of Livestock Development Turkana District.

UNDP Drylands Development Centre (2006), *Making Rural Markets Work for the Poor*, Nairobi, Kenya: UNDP.

United Nations Development Programme (2004, 2007, 2009 editions), *Human Development Report*, New York: UNDP.

Watson D. J. and van Binsbergen J. (2008), *Livelihood Diversification Opportunities for Pastoralists in Turkana, Kenya*, International Livestock Research Institute Research (ILRI) Report 5, Nairobi, Kenya.

6

MYANMAR

Wendy Poussard and Joanna Hayter

*D*r. *Wendy Poussard has been working in international development projects in Asian and Pacific Island countries for more than 30 years and has lived in Fiji, Vietnam, and South Korea. She has conducted research on conflict transformation, peacebuilding, and community-based disaster mitigation and is one of the founders of the International Women's Development Agency (IWDA).*

Joanna Hayter is an international health and development specialist and the executive director of IWDA. She previously was a consultant supporting institutional development and capacity-building initiatives across Asia. During her 25 years in international and community development, she has directed international NGOs in Vietnam and Myanmar and has worked as a program leader, policy adviser, and trainer across 13 other countries in Asia, Africa, and the Pacific.

> When I manage the volunteers I try and boost their confidence. I tell them that they are powerful, they are responsible and they are confident, even though they are poor and share their own small amount. Now they have nothing. Their livelihood systems have been destroyed. Gardens, coconut trees and other crops will take at least five years to get back to the previous standard. They need to repair and replant. People are very humble and not greedy. They also do not want to depend on others.
>
> Anonymous interviewee from Myanmar
> (Centre for Peace and Conflict Studies, 2008: 153)

> It is not enough merely to provide the poor with material assistance. They have to be sufficiently empowered to change their perception of themselves as helpless and ineffectual in an uncaring world.
>
> Daw Aung San Suu Kyi, formerly imprisoned leader
> of the National League for Democracy, Myanmar

CONTEXT

Myanmar[1] (formerly known as Burma), known for its ongoing political conflict, is also familiar with natural disasters. Heavy rainfall means the country is prone to floods and landslides in the monsoon season, and sometimes there have also been earthquakes, cyclones, and droughts. As the largest country in mainland Southeast Asia, it has a population of 49 million fed by its rich rice fields.

Rice is grown on the lowest-lying land in the Irawaddy Delta, in the Ayeyarwady Division, a southern coastal region that was the most affected by Cyclone Nargis in 2008, where land is rarely more than six feet above the sea level. Mangroves, which once provided a buffer against storm surges, have been removed to make way for rice cultivation, which was increased dramatically during the British colonialism when Myanmar was the world's largest rice exporter.

Myanmar also exports natural gas; agricultural, forestry, and fishing products; and gems, but despite its abundance of natural resources, it is categorized by the UN as a "Least Developed Country." About two-thirds of the people in Myanmar live in rural areas and depend on agriculture for their livelihood. Most people in the Irawaddy Delta are farmers and farmworkers; others are fishers, craftsmen, and small service or business workers. Land is wealth, and the poorest people are the ones that are landless, making a living on the margins of the rice paddies. Being wealthy in the Irawaddy Delta means simple things such as having a house of concrete and hardwood rather than bamboo.

Myanmar ranked 132 out of 177 countries on the United Nations Development Programme's (UNDP) Human Development Index (UNDP, 2008: 231) in 2008. A third of the population lives in acute poverty, and the average household spends three-quarters of its income on food, leaving little for shelter, health, and education (UNDP, 2007: 6). Life expectancy at birth is 61 years, about a decade less than all other Southeast Asian countries except Laos. Myanmar has the highest rate of mortality for children under five years in Southeast Asia and the lowest per capita spending on health in the world (WHO World Health Statistics, 2008).

Prior to Cyclone Nargis in 2008, sanctions and other political restrictions limited aid and trade. Burma had the lowest per capita aid in Southeast Asia, at $2.50 per capita a year (compared to Laos $50, Cambodia $35, Bangladesh $6, according to the *UNDP Human Development Report,* 2004: 200). The biggest expansion of aid, prior to Cyclone Nargis, was for HIV/AIDS, and this increased from less than $1 million to $21.5 million in 2005 (Joint United Nations Programme on HIV/AIDS, Myanmar estimate).

Before and after colonialism, Myanmar has been marked by tensions between groups seeking political and economic power. Burmese (known as BaMa) make up 65 percent of the 50 million population and the bulk of the *tatmatdaw* (military) currently in power. Even the name Myanmar itself is contested by ethnic groups, introduced in 1989 (comprised of words *myan*, meaning quick, and *mar*, meaning hardy, and used by BaMa people when referring to their sense of collective identity and culture). Other ethnic groups include the Karen 9 percent, Shan 7 percent, Indian 7 percent, Chin 2 percent, Mon, Kachin, and Wa at 1 percent each (Matthews, 2001: 1–3). Most of the ethnic minorities live in the mountainous border regions that surround central Myanmar.

Ethnic tensions were heightened in the Second World War as Buddhist leaders started their movements toward independence, while Karen leaders allied themselves with the British colonial powers providing labor and even troops for the war effort. Tensions continued after the war, with a Buddhist nationalist leader, General Aung Sun, being assassinated months before independence was declared in 1947.

After a decade of turbulent democracy, in 1962 a coup began the first military regime, which put in place the military control of Myanmar that has lasted until today. Conflict involving the repression of opposition to the military regime, nonviolent resistance by the banned National League for Democracy, led by Nobel Peace Prize–Winner Aung San Suu Kyi and the organization of ethnic minority groups that engage in political activities and sometimes armed fighting for autonomy, continued till the present (Centre for Peace and Conflict Studies, 2008: 14).

A year before Cyclone Nargis, popular protests arose in September 2007 following a hike in fuel prices. Tens of thousands of people took to the streets to express their dissatisfaction with the regime. More than 30 people were killed and more than 2,000 imprisoned (International Crisis Group, 2008: 1).

Traditionally in Myanmar, monks and students have highlighted issues of justice, and the military has sought to curb the freedom of their institutions and leaders—for example university student's parents must sign letters of guarantee that their children will not participate in political activities, and some of the more vocal Buddhist institutions have been shut down by the military (International Crisis Group, 2001: 16–17). In Myanmar most people relate to each other along family patterns with elders viewed as senior and honored with formal communication, so symbolic acts of dissent, such as not bowing to a military official, can be seen as defiant and threatening. The concept of *arnade*, a fear of causing embarrassment or humiliation to another person, strongly influences social interactions.

The 300,000 monks and novices, collectively known as the *Sangha*, at several points over the past two decades have refused to accept alms from military personnel and their families or preside over funerary ceremonies. Without these ceremonies, souls of those who depart are unable to prepare for the next life. When the monks, during the Saffron revolution, marched with their alms bowls turned upside down so that nothing could be put inside, they were both threatening and haunting the military about their regime's refusal to take responsibility for the suffering of the living. As the protests were spearheaded by monks and novices in their bright-orange robes, it became known as the "Saffron revolution."

Meanwhile, the military and other groups often rely on tributes and bribes, black-market goods, logging and mining, and other unregulated businesses (including drugs) to enrich themselves and buy arms (South in Skidmore and Wilson, 2007). Myanmar's military regime has long been scrutinized and criticized by the international community and media for its human rights abuses. International economic and diplomatic sanctions have been applied for much of the past two decades, including restrictions on international aid by bilateral donors and international organizations. The long cycle of conflict and isolation creates human insecurity evident in disparities between regional and ethnic conflicts, rural-urban divides and gender inequality, with poverty, HIV/AIDS, tuberculosis and malaria, malnutrition, and illiteracy particularly high in many parts of the country.

Despite all of this, there are many informal and formal community groups at the local level, which are involved in community activities (such as literary clubs, following the country's long literary tradition including a love of satire). However, community groups must exercise caution as they may be subject to government scrutiny and sanction.

This insecurity means that people in Myanmar tend to rely on family and community networks and organizations rather than the state or more formal institutions. A survey on whether the people trust fellow citizens found that 20 percent of the people in Myanmar said "yes," 23 percent said "no," and 45 percent said that "it depends." At the same time, only about 10 percent did not belong to any civil society organization and about 30 percent to one, over 40 percent to two and about 18 percent to three or more (Steinberg, 2006: 166).

Into this environment of political repression and poverty, came Cyclone Nargis in May 2008. It was first a tropical storm that developed over the ocean with a large system of winds spiraling in toward a region of low-atmospheric pressure. Initially expected to hit Bangladesh, on May 1, 2008, Cyclone Nargis rapidly intensified and reached peak winds of 215 kilometers per hour (135 miles per hour) as it approached Myanmar making it a Category 4 storm. The next day the cyclone reached land in the Ayeyarwady Delta, the rice bowl of Myanmar, with a four-meter wall of water and 200 kilometers per hour winds. These continued to sweep across the country, reaching Yangon city on May 2 and 3, 2008.

Official figures state 84,500 people were killed as a result of the storm, 19,000 were injured, and 2.4 million people were affected. Women and children were more likely to die as a result of the storm, with some villages losing all or most children under seven years of age, and "in some severely affected villages, twice as many women aged 18–60 died as men" (Tripartite Core Group, 2008a: 156). There were 53,800 people missing—a trauma made more significant by the belief that spirits of the dead stay around their bodies for seven days, and unless the family conducts certain funeral rituals they cannot ensure better fates for the dead in their next lives. Cyclone Nargis was soon hailed as the deadliest tropical cyclone since the 1970 Bhola cyclone, which killed almost 500,000 people.

For the region's main source of income—agriculture—much of the end of the harvest that accounts for 25 percent of annual production in the area was gone, as well as several rice warehouses and their stocks. A total of 149,000 buffaloes were lost and 300,173 acres of farmland was flooded (Tripartite Core Group, 2008a: 38). Fishing equipment, fish processing plants, and salt farms were also badly damaged.

In the capital, Yangon, damage was significant—roofs were ripped off, the electricity grid was shut down for the city's 6.5 million people, and the streets were littered with debris. Around 800,000 houses were affected (450,000 totally damaged) as well as over 7,900 factories and commercial buildings.

Three-quarters of health services in the affected areas were damaged along with 4,000 schools. Damage and losses caused by Cyclone Nargis was estimated at $4–4.13 billion.

Research Outline

Due to the sensitive security situation of people who live and work in Myanmar, only a small number of people could be interviewed, and the identity of all people interviewed is concealed. Four interviews were conducted with anonymous local NGO leaders in Myanmar in 2009. Interviewees were selected based on personal contacts and opportunities to talk in a safe environment. These interviews were supplemented by previously published research to bring together experiences and insights of people living and working in Myanmar who were involved in the response of local communities and organizations to Cyclone Nargis. The sources used include the extensive official reporting on post-Nargis recovery prepared by representatives of the Government of the Union of Myanmar, the Association of Southeast Asian Nations (ASEAN), and the UN, with support from the humanitarian and development community. Also included are independent reports and journals of international organizations—particularly a report from the Centre for Peace and Conflict Studies that directly addresses the topic of this book, community resilience. Entitled *Listening to Voices from Inside: Myanmar Civil Society's Response to Cyclone Nargis*, this report documents the experience of 15 local organizations and 32 individuals involved in supporting local communities.

COMMUNITY RESPONSES

When the staff of the international NGOs in Yangon arrived for work on Friday, May 2, 2008, they heard from colleagues or read in emails that a big storm was coming toward Myanmar, but this information had not been transmitted by the mass media. The International Committee of the Red Cross warned the staff to take safety precautions. Offices were closed, and the staff was advised to buy extra drinking water, candles, matches, and enough food for several days. Most people did not react to the warnings, it seems, with any sense of impending disaster.

> Actually when we heard the warnings, we didn't take them very seriously, I remember writing an email to someone overseas and flippantly saying, "Talk soon...if the cyclone doesn't wipe us out!!" We just thought it would be a normal wet-season storm. We all kept working, didn't go home early and didn't prepare for anything. (NGO interviewee, Yangon)

The staff in one office looked up the cyclone on the internet. They saw the number indicating the force of the cyclone but did not understand what that meant and felt overwhelmed by a mix of accurate information and rumors. A false sense of security emerged as the afternoon went by; the skies were still clear, and many believed that the storm had passed:

> In the afternoon I turned on the radio, but they didn't tell us that the epicenter was Myanmar or that it would hit the delta—just that it was coming. At 9.00 P.M. the voltage changed in our house. It was suddenly very bright. I assume this was because the government had turned off power in some areas and so we got the full voltage supply to our house. We also had a telephone call from my Mum's friend who lives in Singapore. He had very exact information and details—then we were nervous. (NGO interviewee, Yangon)

> By 11.00 that night the winds had arrived. Roofs began flapping, heavy rains poured down, and the sky changed color: The red sky—we had never seen anything like this before in our lives. I was at home, worried and scared. We have never experienced a storm like this before. (NGO interviewee, Yangon)

People on the Delta also reported feeling unprepared:

> Before the Cyclone, the radio warned that a storm was going to come and it would be scary like a ghost. But they should have said it would be like a monster or a giant. (NGO interviewee (Centre for Peace and Conflict Studies, 2008: 153))

> I am 60 years of age and I have never experienced this kind of situation in my life. Never have I seen a catastrophe on this scale. I have seen flooding but not this kind of devastation. At a national level, it is difficult to foresee how civil society could be expected to deal with this kind of catastrophe. It was very new to us. (Community leader (Centre for Peace and Conflict Studies, 2008: 169))

People could not find shelter in the storm-swept plain, often climbing trees for protection from rising floodwaters but suffering injuries from the debris underfoot and from above:

> We saw many injuries as a result of Cyclone Nargis. Some had large cuts on their whole bodies from climbing up coconut trees and clinging onto them for up to eight hours. In areas where there were lots of trees and plants, people were safer; they could grab hold of the trees. In the plains, people were more vulnerable. (Health worker (Centre for Peace and Conflict Studies, 2008: 97))

People were immediately concerned with finding family members, with many having lost their family in the chaos of the cyclone. Large numbers of women and children died while trying to find shelter from the storm. Some people were in shock, not speaking, shaking, or hearing voices. The distress was great as people described:

> People lost everything. Some people would be still living with their dead relatives as they had only their hands to bury them and they could not dig the ground. (NGO interviewee (Centre for Peace and Conflict Studies, 2008: 150–151))

From the survivors we heard stories of parents having lost all their children or having to make heartrending choices of which child to save when they could not save all; children who had become orphans and are now living in new homes and towns; people who had been partially submerged in water, not knowing whether they would live or die; both young and old witnessing death and corpses.

> In the weeks after the cyclone, we know of children and adults who are still traumatised by the sound of rain and wind. There are people who are wandering aimlessly and not wanting to start afresh, and children not wanting to talk or play. There are also men and women who have lost their spouses, children who cry themselves to sleep each night and the elderly who have no one to care for them. (Trauma counselor (Centre for Peace and Conflict Studies, 2008: 180))

> For many villagers they had to restart their life and it was a new life. Most of the villages were empty because of Cyclone Nargis. It had killed so many that in some villages only 10–20 persons were still alive. (NGO interviewee (Centre for Peace and Conflict Studies, 2008: 97))

The experience of the cyclone was a stressful time even for those in the city of Yangon who were more likely to have permanent shelter:

> I went into the sitting room and sat on my prayer mat and prayed. I thought of those living in huts and what would be happening for them. I prayed for rain because it seemed when it rained the wind was slower. My cat was so afraid. I thought I wouldn't see my Mum again. . . . I sat down and prayed for people who weren't living in a protected place like me. I prayed for survival, for the storm to stop, prayed for others not myself. (NGO interviewee, Yangon)

> All our windows broke, the noise was deafening. Trees were smashing all the cars. It was really scary. We just stayed in, moved into middle room where there were no windows. It gave us a false sense of security because there was less noise. (NGO interviewee, Yangon)

As morning broke in Yangon, it was difficult to assess the impact of the cyclone because the rain was so heavy. By midday the storm "just stopped," and people came out of their homes and started walking around their neighborhood. Some had heard rumors that the calm they experienced was just the eye of the storm, and that the wind would come again in a few hours.

As it began to clear and we could look out from our apartment window and see the carnage—all the roofs had gone around us, all the oldest grand trees had fallen, trees over 100 years old were gone. (NGO interviewee, Yangon)

It was a terrible thing. Trees, satellite dishes were everywhere, like a nightmare. (NGO interviewee, Yangon)

Neighbours gathered in the streets and worked together to assess the damage. They visited other houses in their lane to ask neighbours if they were all right. Monks and their novices also made house-calls, but some people noted that highranking officials had many soldiers at their gates who did nothing to help. People consistently described a sense that no-one would be coming to help them. Instead cleanup and rescues from community members started immediately: Within an hour a big group of young men with knives started to cut and remove branches— they just came and cleared all our lanes. We didn't know these boys— we had never seen them before. (NGO interviewee, Yangon)

Monks were seen in the streets cleaning up the debris and fallen trees and taking care of lost and orphaned children who followed them, often also helping with the cleanup (*New York Times*, June 18, 2008).

Despite the heavy rains, one of the difficulties of surviving after a cyclone is the lack of clean and fresh water to drink. People reported surviving on drinking coconuts and developing ways of collecting rainwater to drink, and eating any kind of food they could find, often risking their health by eating dead water buffalo and spoiled grain. Cooperative efforts to find and cook food occurred in the villages and Yangon:

The markets were gone so we had no food. We needed to arrange this, but it was very difficult to get to any place to buy food, the roads were blocked everywhere. We could not get through....We set up local supply source for food. We didn't have time to worry about the household level; we had to organize this for everyone, so we arranged for potatoes, rice, beans and oil for many families to be purchased from a local person who could arrange regular supply. (NGO interviewee, Yangon)

In Yangon, the days responding to the cyclone were described as one of cooperation, generosity, and goodwill as people worked together and helped each other:

> Neighbours were all helping each other—it was collective thinking...we didn't just do things for ourselves, we thought of others, so when we went to buy food we bought lots of food for others to share. We pooled money to get electricity supplies back. I kept thinking this could only happen around our area because people are a bit wealthier, but most places would not have been able to do this. Our ability to get money gave us resources more quickly than others.

> Immediately, people offered shelter and accommodation for those who had lost their houses or had so much damage that they had to move their things and sleep somewhere else. This happened very quickly. They needed protection from the weather and safety for their things. My sister was very brave—she went to find the old lady and her daughter whom we had not seen all day, and she pushed through dangerous things to reach them and kept calling and calling until the old lady came out crying, and didn't stop crying. So, we brought her into our house and gave her coffee. Then we all shed tears. (NGO interviewee, Yangon)

> The cyclone proved Burmese people are very capable. If you only have one hundred kyat, and someone else needed that, then you would willingly give it to them. (NGO interviewee, Yangon)

In the Delta, most people went to monasteries for shelter, but camps, spontaneous and government organized, also sprung up. Survivors continued to arrive for many days after the storm at the camps, and people registered at government camps to receive rice. Volunteers from religious organizations and local NGOs were often working at the camps, some themselves affected by the disaster:

> Before helping others, I had to help our house. Our roof flew away. We had to sleep under the sky. There was a lack of water. I had to wait two to three hours for one bucket of water. Two days later, I heard many things about the Delta area. When I heard it was people from my ethnic group that were affected; my blood relatives; I could not stay at home so I went to the camps.

> When I arrived at the camps, I could see that women were menstruating without any sanitary materials. There was a strong smell from this. There were also pregnant women. I knew what they needed so I came back to Yangon and gave 200,000 kyat to buy pads and donate them

to the women. I also collected clothes from relatives and brought them to the Delta. I donated these to the Laputta region. Laputta was the most affected area here. I went back to the camps and helped the project officer. They counted on me for psychosocial and trauma healing. They developed a trauma management program. I contacted people from the camps and we arranged to visit. They had 4,000 victims. I went there around 24/25 May as a logistics person for the trauma healing staff and as a translator and facilitator.

I heard many sad stories from people. I had to take rest to be calm and stable. Every household had lost at least one person. (Trauma counselor (Centre for Peace and Conflict Studies, 2008: 133–134))

Local government–sanctioned organizations like the Red Cross were able to respond quickly—this organization provided shelter to 80,000 households—key supplies such as tarpaulin, ropes, saws, shovels, hoes, nails, so people could start building with local timber—using local volunteers. Other spontaneous efforts of Myanmar people to help were convoys of trucks with drinking water, clothing, and food to take to the Delta:

We spent the next 20 days in the Delta. Over that time many NGOs came and went after they had studied the situation. Local people brought truckloads and truckloads of relief supplies from Yangon and beyond. They brought candles and clothes also. They drove into the Delta area with their supplies and set about distributing it. This was not systematic and was chaotic, but the need was so great. Children were by the roadside waiting and asking for help. (NGO worker (Centre for Peace and Conflict Studies, 2008: 62))

Other efforts included volunteers and funds to supply free funerals and networks of trauma healers and counselors. One group described contacting a network of 200 trauma healers who went out into communities to assist people traumatized by their experiences (Centre for Peace and Conflict Studies, 2008: 25). Existing NGOs reported a huge increase in volunteers; one reporting that numbers went up from 100 to 2,000 in the months following the cyclone. Donations and contributions also were high—a recorded contributions in local cash and kind of $11.86 million by June 24 that year, and the Tripartite Core Group noted that there was probably at least that amount again in unrecorded donations making it difficult to estimate the role of the Myanmar public and civil society in responding to Cyclone Nargis (Tripartite Core Group, 2008c: 40).

Most of these local nongovernment and volunteer efforts were low-profile community endeavors. Others were backed by businesses:

> In the Delta, we brought our mobile medical teams and worked with the health teams in the area and the village. We have been working in the area with our mobile medical teams for a long time so have strong local foundations.

> We copartnered with a Yangon-based businessman for our Nargis relief activities. This businessman is a devout Buddhist and a member of the Board of Trustees of a well-known monastery in Yangon. The village in which we stationed our medical team as a base is the village in which his mother was born. He had also donated the funds required to build the village clinic and we treat and operate on patients in that clinic. The man also hosts our team in his house as most other buildings have been washed away. (Health worker (Centre for Peace and Conflict Studies, 2008: 62–63))

Local businesses contributed goods and donations totaling $68.13 million by June (Tripartite Core Group, 2008c: 41). Other people used their profile inside and outside Myanmar to raise support to help cyclone survivors. Many of these were religious people:

> A religious leader was overseas when the cyclone occurred. He flew back from Australia and started to help. He has a lot of public support so was able to bring a lot of assistance to the affected areas. With this assistance he was able to open a clinic and provide food and shelter to the local people.

> The government knew they couldn't stop this monk. As a result of him a great deal of assistance made it to the affected areas. This monk really led the way and started helping people. This took place whilst other NGOs were still negotiating with the government to be allowed to assist.

> This religious leader was able to generate three million US dollars in two to three weeks. This was the first time he had been involved in relief but he was able to mobilise and provide assistance to more than 500 villages in one township.... In the near future we are also organising a conference. The monks are really influential in Myanmar society and the conference will bring together monks and other social actors from the emergency sector.

> This was his idea. It was also his idea to support the monastic education system in the Delta area....After Cyclone Nargis, many schools were destroyed and the reconstruction of these schools will take a long time. Trust funds have been established by him to support the teachers and the monastic education movement. (Environmental NGO worker (Centre for Peace and Conflict Studies, 2008: 71))

Other organizers were from the arts such as comedian Zarganar (stage name Tweezers), who used comedy to raise funds and send five aid runs to outlying villages in the Delta, about $6,500 of relief a day (*New York Times,* June 6, 2008), but he attracted a high profile and was imprisoned by the government for speaking about the government's response to the natural disaster.

When the international media started reporting what was happening in the camps and by the side of the road as relief was delivered, the government tried to control the survivors of Cyclone Nargis and the local people providing assistance:

> The government tried to force the camps to close and move people to government camps. People were told that if they did not go they would be in trouble. They were worried about not having enough food. As a result of government force, 2,200 people had to move to the new camp. They would come back to old camps for food and other things. (NGO interviewee (Centre for Peace and Conflict Studies, 2008: 134))

> The government began checking with people why they had brought these things. The government wanted to know which organisation they were from and where they got the money from, etc. This made the people giving donations feel bad and stop coming to the Delta. At the same time, the government would stop people coming to the road to collect relief supplies. Gradually the vehicles reduced. (NGO interviewee (Centre for Peace and Conflict Studies, 2008: 151–152))

Local groups, some practical in nature focused on community activities, and others mixing needs (shelter, clothing, food) with social and psychosocial support networks and activities (such as listening and sharing stories) sprang up all over the Delta looking for funding for recovery. The *Post-Nargis Joint Assessment* reported that half of all villages had self-reliance groups, one-third women's groups, and one-fourth religious associations.

Athauk Apun (ATP) was one group facilitating small grants to what they called "self-help groups," groups of neighbors and community members actively responding to help themselves and others. With small grants ($5,000 or less to communities) to buy food and shelter and later rice seed, diesel, boats, fishing hooks, lines, and net, communities were able to start to recover:

> Many self-help groups described the ATP approach as a motivation for local action. In some of the more traumatized villages, there were alarming signs of lethargy and depression, and several survivors spoke openly of being close to suicide. Seeing or hearing about the activities

of groups that had accessed ATP grants seems to have had an important galvanizing effect. (ATP, 2008: 24)

An estimated 350,000 survivors received emergency aid via self-help
groups funded through ATP. For the Local Resource Centre (LRC),
small grants directly to local groups were key to helping with the
sense of powerlessness people felt after the disaster. In the words of
aid workers Hedlund and Myint Su (2008: 20): "Not only saving
lives, these grants were responsive, cost-efficient and some argue cata-
lytic with regard to promoting dignity, self-reliance and recovery."

Local NGOs were the first outsiders to reach people in cyclone-
devastated areas and deliver relief programs. They began with volunteers and then became more systematic as more and more people
arrived and got organized, and international networks that had existing partnerships with local NGOs were able to respond quickly to
people's needs.

A critical factor was people's ability to utilize personal contacts and
build networks that systematically linked with decision-making structures. Lack of established structures and systems sometimes proved
an advantage. Individual private providers of aid were more likely to
consult with villagers and village committees on needs and priorities
and to provide directly to villagers. They targeted fewer villages and
so had more time for consultations because, as informal aid providers, they were not constrained by sector or organizational guidelines
(Tri-Partite Core Group, November 2008: 14). After Cyclone Nargis,
it was a time of organizational growth and collaboration:

> We were [a] very small NGO at this time, just beginning: We only
> had four staff. We knew from listening to people who had come up
> from the Delta that they were heroes for bringing these stories, but
> we knew we had to do something too. But (our organization) had
> no money. So (a staff member) went to the embassy and they gave
> him $30,000 immediately. We changed it into kyat and then we went
> immediately down to the Delta. Two went east and two went west. It
> was most important to just reach the people and help. (NGO intervie-
> wee, Yangon)

> Some international NGOs came up with [the] idea for a centre to
> provide support to local groups. One organisation provided the space,
> and the Local Resource Centre was born. People said, "When will
> the LRC open?" So we said, "Tomorrow!" They asked, "What time?"
> So we said, "Eleven am!" And it did. It opened on May 15. Lots
> of INGOs [international NGOs] were really great and immediately
> seconded staff. We really appreciated their help. We began to facilitate

grants and funding. It was soon overwhelming. We worked nonstop for twelve hours a day every day for a month. There was no division of roles and responsibilities; we were one team and everyone had to do everything, depending on who came in from the villages and what their needs were. At the LRC we moved through different stages of response. From money plus supplies (tarps, medicines, jerry cans, body bags) to practical training: how to make water clean, how to handle dead bodies, how to plan in an emergency. We advised donors that local groups were doing very well and they could fund them. Because of LRC a lot of partnerships emerged between international and local NGOs. (NGO interviewee, Yangon)

International organizations were able to provide many resources, but if they gave money it had to be distributed through the government. It was the many small organizations who could give financial support directly to families so that they could buy the things they needed. As well as rice, oil, and food, they needed other things like blankets and mosquito nets and clothes. In some areas, however, it was not helpful to distribute money because people could not reach the market to buy things, and what people wanted was food.

Local organizations were pressed for time trying to save lives and care for their staff and volunteers, in a fragile security context. Sustaining motivation and energy levels amongst the staff was a key challenge. Debriefing and storytelling sessions were important ways to maintain commitment and motivation. Celebrating small successes and training sessions on topics such as trauma healing and psychosocial support were also helpful in keeping the staff motivated and maintaining their self-confidence.

Although the delay of several weeks in the international response to the cyclone caused additional suffering, the catastrophic "second wave" of deaths from contaminated water and disease, which had been feared, did not occur. Commentators paid tribute to the achievements of civil society in Myanmar for the way they had responded to the disaster, such as international organization workers, and aid workers Julie Belanger and Richard Horsey (2008: 4) noted:

[The avoidance of many more deaths] is no doubt partly down to luck, but is also attributable to the resilience of the communities affected and the strength of social networks, the extraordinary efforts of local civil society and private donors, the rapid mobilisation of local staff from across the country by agencies already on the ground and the government's own response, the scale and impact of which have not always been fully recognized.

Although Cyclone Nargis caused great destruction and trauma, people's response to the disaster also brought positive and long-term changes for the strength and capacity of communities. Because of Myanmar's isolation from international contact, and the restrictions on civil society imposed by government policy, there has been a perception internationally that the local organizations are not active. The demonstrations of civil society activity in the wake of Cyclone Nargis was seen to debunk this myth and affirm confidence of local people in themselves:

> It is a precious thing—Burmese people like me realized that civil society is powerful. The space for civil society has been recognised, almost respected in some ways. (NGO interviewee, Yangon)

> These opportunities don't come around very often. If we didn't have Nargis, then the government wouldn't listen to us. Nargis is a sad story. It affected many people's lives and people are still suffering. The good side to Nargis was the larger volume of problems it caused. The government has to be more flexible and has to listen to us. They realise that by themselves they can't cope with the problems. We can't afford to lose that leverage for civil society in the country to make more out of this for the future. Many people complain that there is no civil society in Myanmar. When Nargis happened, however, we could prove that there has been, and there still is, a civil society here. We have been saying exactly that for the last ten years. (Local NGO worker (Centre for Peace and Conflict Studies, 2008: 87))

In fact the response of the local people to the cyclone strengthened communities by linking groups of people who had no previous experience of working together in new relationships. The response to Cyclone Nargis brought together communities, business people, religious organizations, government authorities, and diverse community organizations, so people formed new networks and alliances that crossed the divisions of ethnicity and religion:

> After Nargis we found many coordinators for the relief work from the community itself. We were able to establish relationships with the most remote areas, like Hygnie Island. We previously had no contact with these communities. We could coordinate relief with them and it was a way for us to connect and support them. (NGO interviewee (Centre for Peace and Conflict Studies, 2008: 24))

> Our strategy was to involve alumni youth groups and also other youth groups to carry out Cyclone relief work after Nargis.... Before Nargis we did not have any relationships with the local youth groups. This is our new partner in the community. We got to know them during

Cyclone Nargis and we worked with them because they knew a lot about the local situation. (NGO interviewee (Centre for Peace and Conflict Studies, 2008: 24))

We have a lot more linkages with INGOs since Nargis. We share our experiences, skills and knowledge with them. This is a good way of ensuring sustainability. It is also wise because we can't trust that this money will always be provided because of changes in the economy. We also know that for sustainability it is important to build the capacity of women. They can then take up the opportunity. (Environmental NGO worker (Centre for Peace and Conflict Studies, 2008: 73))

The crisis of the cyclone also created space for political changes, which impacted positively on the local peoples' lives:

We have been able to generate a greater influence as a result of Cyclone Nargis. For example, after Nargis the price of diesel really rose. We organised with the private sector to flood the market without profit. The government then also took off the government tax. This broke a government monopoly of ten years. We have been trying to advocate for this tax policy for decade, and Nargis provided us the opportunity to break through it. (NGO interviewee (Centre for Peace and Conflict Studies, 2008: 24–25))

Communities in the Delta and the city of Yangon clearly demonstrated resilience and capacity to organize at a local and national level to respond to the disaster. The official post-cyclone assessment also underlined the role of community in the Irawaddy Delta reporting that it had high levels of social capital due to scarcity of resources, lack of welfare or state safety net, and traditions of reciprocity (Tripartite Core Group, 2008c: 2–3).

With the media profile of Burma and then the disaster, the international attention and response to Cyclone Nargis was significant. The *New York Times* (on October 21, 2008, five months after international aid began flowing to Myanmar) reported that international donors so far provided $100 for every person that survived Nargis, and there were 26 international aid organizations working in one district of the Delta alone.

In some ways, the international media attention and advocacy that decried the human rights record of the government of Myanmar had unexpected negative consequences, as it had denied both the general public in other countries donating money for Cyclone Nargis victims and the aid agencies the chance to see what local organizations and capacity existed in Myanmar, as summarized by the Centre for Peace and Conflict Studies (2008: vi): "Recognition of the space available to civil society in Myanmar goes against external perceptions of what is possible. These perceptions are the result of skewed reportage and the dominance of the voice of advocates for political change in Myanmar who are located outside the country."

International staff, when they arrived, had pressure from their organizations to act quickly and demonstrate what their agency was doing with the donations of people from their home countries. However, their efforts were slowed down when they faced the same challenges that had confronted local organizations such as lack of telephone communications and transport. Some with resources already in Myanmar diverted trucks, aid, and equipment from the north (where conflict exists between ethnic groups) to cyclone-affected areas.

Language was another key problem for international staff dealing with a region that was rural and had few foreign language speakers. In general the international community was isolated in the UN compound in Yangon where few local NGOs had access. Meetings, minutes, and technical and strategic documents coordinating the international response were almost all in English and translations were rare. There was some frustration with the focus on situation assessments by the international actors and disputes between the outsiders and the locals who had been working on the disaster response for the previous month about how to do relief and recovery work.

One interviewee described that the way UN meetings were held meant that local people could not participate:

> The location of the Cluster meetings was also an issue. They are held at a grand hotel. The participants of the women's organisations are from the villages in remote areas. They are worried about their dress and footwear. They have to pay at least 2,000–3,000 kyat to get to the location of the hotel. These people are performing their work on a volunteer basis. They do not have administration funds to cover transport costs. Some have to travel very far to attend the meetings and we provide some travel allowance for them to come. (NGO interviewee, (Centre for Peace and Conflict Studies, 2008: 122–123))

There were other perceived problems with the approach of the international aid agencies. One was mistrust and a perception that the international aid agencies were working for their own interests rather than that of the local groups:

> We need to empower Burmese and locals to do this themselves. To some extent, it's not strange that the Burmese do not trust: They see

expatriates being brought in on high salaries and having airfares provided. They (local people) are lacking capacity also. Capacity-building makes sense and is an investment in the future. (NGO interviewee (Centre for Peace and Conflict Studies, 2008: 86))

There are many local organisations we can work with but we can also work with international organisations if they can be found. We find INGOs want to work by themselves. They have their own rules and principles. It is the goal for them to be able to cooperate and collaborate. They can carry out this work for the long term if they can partner with trustworthy local organisations. Otherwise they affect the situation in different ways.

To reach out to the community, we need local groups. If INGOs come to Myanmar and do not have any contacts, they will set up an office and call for people to work for them who have fluent English and have pursued higher education overseas. People from the city will get the positions they offer and these people will not have connections with local people. This is particularly important in Myanmar where the situation is very difficult and you have to deal with the authorities at every corner. (NGO interviewee (Centre for Peace and Conflict Studies, 2008: 100))

We are very principled in our donor selection. We need donors to have a good knowledge of our reality. We don't want to pretend with them. They must accept our reality and think about our capacity-building. Trust is really important. We want a donor to always be involved with us and working collaboratively. Good or bad, we have shared responsibility for our work. We want to develop proposals collectively and be flexible and open. It can be difficult to find donors like this. (NGO interviewee (Centre for Peace and Conflict Studies, 2008: 123))

We contacted the international community for assistance as we already had connections and the experience of working with them. We knew we needed to initiate contact with them as they have the power. We found their limitations difficult to deal with, however. It was difficult to get the figures they wanted as people would move from village to village. We would predict 3,000 people, but then the figure would be different. (NGO interviewee (Centre for Peace and Conflict Studies, 2008: 151))

For Hedlund and Myint Su (2008: 21), there was a missed opportunity by the international NGOs to back the efforts of local civil society: "Between the self-help community groups and the registered Myanmar NGOs, hundreds of smaller civil society groups sprang into life. All mobilized their own resources and some were partially funded by small grants programs. Many had the potential to do much more. However...at a meeting of donors and INGOs in early June it was announced that 30 such groups had been identified and were looking for partners and funding. There was silence in the room."

International organizations, locals said, should be careful not to disrupt or duplicate the work of existing organizations, and they should commit themselves to work in partnership and strengthen and support local groups, systems, and structures. There was also a strong desire that the focus of external organizations should be capacity building of local organizations (Centre for Peace and Conflict Studies, 2008: 26).

ATP (2008: 25), funder of many self-help groups, reported that the international agencies, with their focus on assessments and tracking aid outcomes, lost sight of the real needs of local people: "What the victims of Nargis really needed was for us to trust them—to trust them as the decision-makers and implementers of their own relief and recovery response."

CONFLICT, PEACE, COMMUNITY, AND DISASTER

The scale and force of Cyclone Nargis created distress and sadness. One anonymous interviewee described a village where the only survivors were men over 40. These survivors felt embarrassed and guilty that they had survived and that they could not save their wives and families. They carved the date of the cyclone on the trees, and tattooed it on their arms and bodies.

Compounding this trauma, Cyclone Nargis also brought to the fore conflict over land. People can live on land if they work it, and with vast areas of farmland being destroyed and expectations of family members over entitlements to deceased people's land, conflicts arose over land within communities. The need to work the land led to debts while the landless became more desperate to find labor on farms.

Cyclone Nargis also highlighted the ongoing conflict between the people and the military with reports of fear of the authorities, imprisonment of some community organizers who criticized the government's response to Nargis, forced labor and child labor directed by the government to rebuild facilities destroyed by winds and flooding, and government favoritism and exclusion in distribution of supplies (Suwanvanichkij et al., 2009: 11).

As local NGOs came into the Delta and self-help groups began to spring into action, the response of local authorities varied. In some areas government officials were seen to be obstructive or self-seeking, in other cases they were helpful and generous:

> The challenge is with organizing groups without letting the government know. We are visible, have tons of stuff. I was very scared for my people [going to deliver aid]. We had to make relations with the government, let them know we are not [political]. . . . We choose villages that have no government office, it makes our work easier. Big villages have a police station, they sometimes want to see permissions. . . sometimes, the headmen are afraid to lose their job and they inform police; foreigners are not allowed. The government is always reporting you, it is unnecessary. . . they tell you can do this, you can't do that. (Relief worker, working in Bogale (Suwanvanichkij et al., 2009: 30))

The majority of civil society groups reported working with the government and working with communities and described the delicate balance that was needed to keep relationships with both:

> There is a demonstrable need for a strong commitment on our part, the implementers, to change our old and strongly held paradigm of generalising authorities as "impossible" and "hopeless" in cooperating with our work and mission. We have proven that once trust and relationship is built, they support our mission even if it means making our work "internally allowed" by them.
>
> Instead of blaming the government, we just worked with them as much as we could. We worked with them and worked with smiles.
>
> While working on building friendship and trust with the local authorities, there is equal effort needed to build trust with communities who normally suspect/perceive authorities as abusive. If we become friendly with authorities, there is a tendency for us to be suspected by communities as favouring the authorities, thus they should be our primary allies in our work to achieve our objectives and goals. (NGO interviewees (Centre for Peace and Conflict Studies, 2008: 27–28))

Local organizations engaged in peacebuilding activities participated in relief activities often for the first time after Cyclone Nargis. They noted the potential created by the crisis of the disaster for different ethnic groups to work together, even in environments formerly marked by conflict and suspicion:

> We didn't care if people were Karen or Buddhist or Burmese, we made no distinction. Before Nargis there was a lot of discrimination between

religious groups, especially Christian, Muslim and Buddhist. When Nargis happened, no one could stay in the house. Everyone came out to work for those affected by the cyclone and people didn't want to stay alone. They didn't see religion and race, they just saw human beings. Those things came out, that spirit. (NGO interviewee (Centre for Peace and Conflict Studies, 2008: 14))

They went to Buddhist and Christian communities and people would not ask what religion they were when receiving assistance. The response was across all religious lines. This was one of the first times that something has happened which has caused a breaking down of barriers between religions.

When I worked there, our religious organisation would very much work in its own area and assert its position as being different from other denominations and religions. The Nargis response went beyond all that. In our experience, Cyclone Nargis really brought people together. People began to form groups to respond to the disaster and that amazed us. (NGO interviewee (Centre for Peace and Conflict Studies, 2008: 34))

Most people we went to interview were Karen people. They said they don't want to work with Burmese people. They say Burmese people are rude and oppress them. This conflict began decades ago. They see this as an opportunity to fight against them. For the Burma side, they said this is punishment to the Karen from God because of the sins they have committed.

When you go to help, people ask you where you are from. When I say Rakhine State, people remembered that our state has a lot of cyclones. They said we will remember you helped us and we will remember you next time. We appreciated that and we cried together a lot. (NGO interviewee (Centre for Peace and Conflict Studies, 2008: 136))

Many local organizations described the experience of responding to Cyclone Nargis as creating new networks, relationships, and trust between different groups:

The situation in Myanmar is confusing and we operate in a context where there are a lot of things we may not understand and there is a lot of emotion. Rumours make people hesitant and fearful. We would hear of other people getting arrested and become concerned about our own situation. We had to find ways of addressing these concerns and carrying out the work. We found building relationships was the key to feeling safe. (NGO interviewee (Centre for Peace and Conflict Studies, 2008: 22))

We learned that emergency projects can open up the way for peace-building to start in areas that had previously been difficult for us to access. With the emergency and relief work as a shared interest among communities and leaders, cooperation and solidarity becomes possible or is facilitated by the emergency efforts as an entry point.

The relationship already established with participants can pave the way for peacebuilding and other program areas to begin to complement emergency components. The relief phase is all about needs and addressing them rapidly. However, as we move into the rehabilitation stage it's about building communities. We need to work collaboratively and sustainably. (NGO interviewee (Centre for Peace and Conflict Studies, 2008: 23))

However, this kind of relationship building was low key, and sometimes forgotten by those not engaged in local-level work and focused on the government level:

The UN described our government's response to Cyclone Nargis as slow. The UN should be on the people's side. The response was not slow, it was silent. I don't understand why the UN didn't say this. This is a negative point about the UN. (NGO interviewee (Centre for Peace and Conflict Studies, 2008: 152))

Local groups found ways to work past obstacles of conflict between different ethnic and religious groups and with the military regime by using the crisis created by Cyclone Nargis to interact with each other cooperatively and by taking a longer-term view, as described by one interviewee (Centre for Peace and Conflict Studies, 2008: 66):

Isolated people are denied exposure to contemporary ideas and approaches. The more they are isolated the more they are entrenched in their own world or comfort zone where they do not have to confront or deal with others. In a divided community we need to learn how to stay together, work together, and develop together without compromising our values and identity. We need to be restored as a peaceful community.

Box 6.1   How a Closed State Opened to Aid

After the cyclone, the Myanmar government declared a national emergency and convened its Natural Disaster Preparedness Central Committee. It called for external assistance but specified a preference for bilateral aid without foreign personnel. Supplies began to arrive and were distributed by the government. Myanmar's international policy had long been guided by an isolationist doctrine of "self-reliance," and the government was reluctant to provide access to Western personnel from international agencies.

Media brought stories of aid supplies waiting in warehouses to be delivered once government access was granted, aid agency personnel bemoaning the delay and restrictions to do their work, and human rights organizations used the opportunity to depict the Myanmar regime as uncaring and repressive.

As it became clear that the authorities would not immediately provide access for the international agencies, the United Kingdom, the United States, and several member countries of the EU applied diplomatic pressure through the UN. France suggested that the UN Security Council should consider militarily intervention, invoking the UN doctrine of "responsibility to protect" under the international law. At the same time, French, British, and American warships carrying aid supplies took up positions near the borders of Myanmar's territorial waters.

Because the government of Myanmar had scheduled its national referendum to change the constitution on May 10, the cyclone came at a politically sensitive time. Despite the devastation, the referendum took place as planned in most areas. The government announced that 92.4 percent of the 22 million eligible voters had voted "yes." Such a landslide approval could not be affected by subsequent voting, but the referendum went ahead on May 24 in the remaining areas worst affected by the cyclone. The actions of the government in delaying international aid and pushing ahead with the referendum brought severe criticism from the international organizations, governments, and the media, exacerbating the adversarial relationship between Myanmar and other national governments.

The breakthrough in facilitating an international response was initiated by ASEAN who, despite the government's refusal to permit foreign personnel, agreed on May 8 to coordinate with an assessment team working through the framework of ASEAN.

On May 19, a special Foreign Ministers Meeting established the ASEAN Humanitarian Task Force for the victims of Cyclone Nargis. At an ASEAN-UN conference in Yangon, international donors pledged their support, and the Task Force formed a new body, the Tripartite Core Group (TCG) consisting of the Myanmar government, ASEAN, and the UN to coordinate relief efforts. At the Pledging Conference, donors made two major demands, access for aid workers and the preparation of an objective and credible needs assessment, which became the responsibility of the TCG.

This ASEAN-led coalition facilitating international assistance was acceptable to the Myanmar leadership, and a meeting between the UN General Secretary Ban Ki-moon and General Than Shwe in Naypyitaw on May 23 resulted at last in a commitment to allow international access to areas affected by the cyclone.

Meanwhile, people from Myanmar were clearing streets, roads, and damaged schools and places of worship, organizing food drives and donations and collective kitchens, setting up temporary shelter and building back houses, providing makeshift health services, establishing psychosocial support groups, and religious leaders were key in providing counseling and support for those who had lost their loved ones.

Box 6.2   Recommendations for Interventions

These recommendations are taken from a key report Listening to Voices from the Inside (Centre for Peace and Conflict Studies, 2008: 45) that draws on interviews with local civil society groups in Myanmar following Cyclone Nargis and place a particular emphasis on working with a military regime and oppressed and marginalized communities.

Understand and Work with Local Actors

- Understand the local context (including the social and conflict dynamics, the religious and ethno-cultural diversity of the country, and its political history).
- Be flexible and adjust expectations to fit the environment.
- Be creative (including using informal ways to overcome obstacles with access to communities).
- Work with the local groups and work with different groups (including avoiding overlap of what the local organizations are doing or changing existing power structures).
- Work with the government (including recognizing that the government is part of the local reality and needs inclusion to develop and improve people's lives).

Build Local Capacity

- Trust the local people and strengthen local capacities.
- Build the capacity of the local people (including individuals and organizations responding to disasters).
- Focus on young people (to foster future development of the country).

NOTE

1. The country was formerly known as Burma since British colonial rule. In 1989, the military government changed the English version of the

country's name from "Burma" to "Myanmar," as well as other anglicized names such as changing the former capital city from "Rangoon" to "Yangon." The United Nations endorsed the change, and officially the state is referred to as the Union of Myanmar.

REFERENCES

ATP Staff (December 2008), "Helping the Heroes: Practical Lessons from An Attempt to Support a Civil Society Emergency Response after Nargis," *Humanitarian Exchange*: 23–25.

Belanger, J. and Horsey R. (December 2008), "Negotiating Humanitarian Access to Cyclone-Affected Areas of Myanmar: A Review," *Humanitarian Exchange*: 2–5.

"Burmese Endure in Spite of Junta, Aid Workers Say," *New York Times*, June 18, 2008.

Centre for Peace and Conflict Studies (2008), *Listening to Voices from Inside: Myanmar Civil Society's Response to Cyclone Nargis*, Myanmar.

Dapice, D., Vallely, T., and Wilkinson, B. (2009), *Assessment of the Myanmar Agricultural Economy*, Ash Institute, International Development Enterprises (IDE) and Ministry of Agriculture and Irrigation of the Union of Myanmar.

Hedlund, K. and Myint Su D. (December 2008), "Support to Local Initiatives in the Nargis Response: A Fringe versus Mainstream Approach," *Humanitarian Exchange*: 19–22.

Humphries, P. (December 2008), "Nargis and beyond: A Choice between Sensationalism and Politicised Inaction?," *Humanitarian Exchange*: 10–13.

International Crisis Group (2001), *Myanmar: The Role of Civil Society*, Asia Report No. 27.

International Crisis Group (2008), *Myanmar after Nargis: Time to Normalise Aid Relations*, Asia Report No. 161.

International Crisis Group (December 2008), *Conflict History: Myanmar/ Burma*.

Matthews, B. (2001), *Ethnic and Religious Diversity: Myanmar's Unfolding Nemesis*, Visiting Researchers Series No. 3, Institute of Southeast Asian Studies.

"Post-Cyclone Aid Divides Myanmar Between the Helped and the Helpless," *New York Times*, October 21, 2008.

Skidmore, M. and Wilson T. (eds.) (2007), *Myanmar: The State, Community and the Environment*, Canberra, Australia: Asia Pacific Press / Australian National University.

Steinberg, D. I. (2006), "Civil Society and Legitimacy," in T. Wilson (ed), *Myanmar's Long Road to Reconciliation*, Australia: APP.

Suwanvanichkij, S., Mahn M., Maung C., Daniels B., Murakami N., Wirtz A., and Beyer C. (March 2009), *After the Storm: Voices from the Delta*,

Emergency Assistance Team Burma / John Hopkins Bloomberg School of Public Health.

Suu Kyi, A. S. (1994), *Empowerment for a Culture of Peace Development*, Address to a meeting of the World Commission on Culture and Development, Manila, November 21, 1994.

Tri-Partite Core Group—Union of Myanmar, Association of Southeast Asian Nations, United Nations, (November 2008), *Post Nargis Social Impacts Monitoring*.

Tripartite Core Group—Union of Myanmar, Association of Southeast Asian Nations, United Nations, (2008a), *Post-Nargis Joint Assessment*.

Tripartite Core Group—Union of Myanmar, Association of Southeast Asian Nations, United Nations, (2008b), *Post-Nargis Recovery and Preparedness Plan*.

Tripartite Core Group—Union of Myanmar, Association of Southeast Asian Nations, United Nations, (2008c), *Post-Nargis Periodic Review*.

United Nations Development Programme (2004, 2008), *Human Development Report 2007/2008*, New York: United Nations.

United Nations Development Programme / UNOPS / Ministry of National Planning and Economic Development (June 2007), *Integrated Household Living Conditions Survey*, Yangon: Union of Myanmar.

Wilson, T. (ed.) (2006), *Myanmar's Long Road to Reconciliation*, Singapore: Asia Pacific Press Institute of Southeast Asian Studies.

7

CONCLUSION: COMMUNITY
RESILIENCE IN NATURAL
DISASTERS

Looking back on the experience, the story was the modern age's typical natural disaster. The causes were part natural (the disaster) and part human (poor infrastructure and planning) in the large city. When the disaster struck, it was the poor people living in the most vulnerable low-lying areas who were most likely to lose everything and least likely to have the resources to escape. When people responded, it was as groups of families trying to save themselves, then groups of volunteers with transport who went out to rescue strangers. But the government, army, and outside agencies took actions to impose control, not resilience, in the local population—they restricted movement, created relief camps, and then left people waiting and dependent.

When people found out about the huge scale of the disaster, volunteerism was similarly enormous, unprecedented, spontaneous and organized, religious, and civil. However, the government focused on its staff responding to the disaster as the ones in charge, the heroes, the experts. The media from outside the city called the poor people fleeing "refugees" and depicted them as helpless. Rumors about "looting" and lawlessness were left to run rife in news and commentary, leading television viewers to be confused about whether or not to donate. The media from inside the city worked 24/7 to report on the voices of survivors in the city and to get information to those who were left wondering about what was happening—breaking all precedents on their readership and listenership. Graffiti and images that depicted the identity of the city as wry, tough, and eloquent became huge hits on social-networking sites, as people sought to reconnect with the city that they had had so many good times in.

This city was New Orleans while experiencing the force of Hurricane Katrina and the aftermath of flooding. Despite being a developed and powerful country, what happened in the US state of Louisiana in 2005 could have equally happened in any of the places described in this book. Parts of the story of the people who survived Hurricane Katrina will be familiar to the people we interviewed—from the desert of Kenya to the mountains of Pakistan.

The influence of the media and official pronouncements is such that people believed that New Orleans was a mess of crime, poverty, and social problems in the aftermath of Hurricane Katrina. Of course, these elements were there, but the major element was the extraordinary stories of resilience. To give just a few examples, documented by the historian Professor Douglas Brinkley from Tulane University in Louisiana, in his chronicle of the disaster:

* Groups of people with boats spontaneously rescued people from the floodwaters, for example the so-called Cajun Navy of 35 small boats rescued 4,000 people (Brinkley, 2006: 381).

* Usually people stranded on rooftops were not panicked but well-organised by self-appointed leaders who kept people calm and focused on survival, as Coast Guard Lieutenant Chris Huberty described: "I can't tell you how many times a man would stay behind an extra day or two on the roof and let his wife and kids go first. It broke my heart. We'd go to an apartment building and you'd see that someone was in charge, organising the survivors. We'd tell him, 'We can only take five,' and they'd sort out the worst cases. It happened many times that the guy in charge was the last to leave" (Brinkley, 2006: 328).

* A charity hospital known locally as "Big Charity," the hospital for the poor, continued to run despite the floodwaters, lack of electricity and other obstacles, with doctors and nurses working to do what they could for their ICU patients.

* Support to rebuild New Orlean's famous artistic community after the disaster was enormous—for example a new organization the New Orleans Musicians Relief Fund assisted more than 775 musicians to purchase instruments and in some cases pay rent (Brinkley, 2006: 253).

As for the Superdome, a place where supposedly thousands of people existed in an anarchic state, despite living with terrible sanitation and frequent hyperventilation due to the lack of air in the dome, there were only six deaths from natural causes, one person died of an over-dose and another committed suicide. Instead of chaos, people were remarkably cooperative with each other and orderly, as National Guard Colonel Thomas Beron said: "Don't get me wrong, bad things hap-pened, but I didn't see any killing and raping and cutting of throats or anything.... Ninety-nine percent of the people in the Dome were very well behaved" (Brinkley, 2006: 193).

Television reporting of Hurricane Katrina is a particularly dramatic example of when local community resilience was not just underac-knowledged but subverted through reporting that overemphasized rumors and isolated incidents. However, this "frame" of envisaging community responses to natural disasters has had an enormous impact on the way many people look at natural disasters. We have been trained by the Western media and government agencies to look for social problems after disasters, the *unusual story*, not community resilience, which is the *usual story*.

When we started out with this book, our intent was to look at many elements of community resilience, and one question on our minds was—why did community resilience appear to happen in some places and not in others? Instead of an answer to this question, we found out that the question had been biased by the "frames" we learned from our governments and media. Despite the wide variation in the different contexts, from the rural communities in Kenya to one of the world's largest urban areas, Mexico City—community resilience was found *everywhere*.

Perhaps we would not have found resilience if we had not done interviews with the local people. Despite being transcribed into English, the interviews retain a local flavor that is evocative of the place and culture from which they are drawn. There is, for example, the rapid to-and-fro of a friendly chat in Indonesia, a quiet conversa-tion under the trees in Kenya, where one can almost see the heads nod as a point of agreement and understanding is marked by "ya," and the thoughtful silence that is the appropriate response to the reflections of a religious thinker in Pakistan. This diversity in the people we talked to makes the common themes that emerge even more compelling.

They talk about experiences that may be hard for others to even imagine, unprecedented events of overwhelming dramatic intensity. These were experiences that were beyond any similar kind of disaster people had been through before—in Myanmar the sky was red, in Solomon Islands the waves were huge, in Pakistan the earthquake was the strongest people had ever felt. And to communicate the scope of the disaster there is recourse to the surreal language of legends, dreams, movies: Daudy knew of legends of "water that reaches the sky" in Aceh, and when the waves came he thought it seemed like a scary monster: "Oh, that isn't like water...that is like an animal or something, a creature that is very tall, stands sky-high and can destroy everything." The warnings did not capture this language; in Myanmar, an interviewee described radio as telling people it would be like a ghost, but it was a monster, a giant. People referred to biblical and Koranic stories—doomsday, angels, the end of the world.

To look closely at what local people were saying about these extraordinary experiences, we compared them using NVivo, which is a computer application designed to assist in qualitative analysis. First the structure of the coding categories (what we were going to search for) was derived from an in-depth analysis of the Pakistan transcripts. Then the chapters were entered as cases and further codes were added to and modified during the analysis to provide a better fit with the comparative data.

This analysis pointed more to similarities than differences—in each of the cases we looked at, there was no effective warning. In most cases, it meant people had to "think on their feet," or in the case of Turkana's drought it meant being born into a situation of disaster.

Everywhere, when faced with disaster, people have agency. Far from being inert victims, people acted—they ran, they worked, they tried to figure out what was happening, communicated with others, and sought out more information. The second sign of resilience seen in all places was cooperation—most of the stories are about gathering up the children, reassuring elderly relatives, seeking out friends, and coming to the aid of others. Some local people were stunned and needed prodding or direction, but the communities as a whole responded in a much more resilient fashion, taking care of self and family, neighbors, and others. In fact collectively, there was a sort of informal community governance that led the people not coping into following the rest of the community—in Aceh people too stunned to run were told to run, in Pakistan teachers took care of children who fainted, in Myanmar "within an hour" young men with knives were clearing the streets, in Kenya community leaders urged people to find new ways to survive the drought. Everywhere, the first people who came to help were neighbors, relatives from nearby towns, friends from capital cities, and NGOs that are inside the country. They are welcomed with open arms as being not-so-distant friends that arrive quickly to help and have some understanding of the local culture and context.

There was not a scramble of every man for himself, and some paid a high price for their altruism, such as the men in Aceh who went out into the sea after the first wave to rescue people and were washed away, or schoolteachers in Pakistan who went back into buildings to rescue others and died. Volunteering by people sometimes seriously affected by the disaster and at other times by those who fared better in neighboring areas also points to high levels of altruism around the world.

From the initial onset of a natural disaster, people responded collectively to the danger and communicated with each other to try and gain an accurate picture of what was happening and to work out how best to respond. In this sense, talking and understanding what was being said was also very important, even at the point where the disaster might be seen as primarily a sensory and immediate experience. Communication was important to determine things such as: Why are people running? What are they running from? What are they saying? What do they mean by sea water is coming up the hill?

Some reported that they knew how to respond, others said that they lacked information about how to respond, and yet others said that they had conflicting information. Media of one sort or another (television, radio, Internet) was an important source of information more in the aftermath of the disaster than before or during in all places. Only in Pakistan did people get a lot of information about the disaster through watching television. The rest tended to get information through talking to friends.

When recalling the natural disaster, people described the identity of a community as a source of strength. Local identities usually included toughness and capacity. Interviewees in Myanmar's civil society pointed out that the crisis actually built a sense of capacity: "The cyclone proved that Burmese people are really capable." Or we heard Edy, the street children worker, tell us that the Acehnese have a "strong mentality to handle challenging situations. We have those capacities. We used to find ways to survive. Anything. We do everything to survive. All jobs." Or the headmaster Daniel commenting on the innate strength of the Turkana people: "The Turkana may look poor, the way they dress, but they have a very strong spirit, you know, they can move on. I always feel if you bought people from the other parts of Kenya here, they cannot survive a week if they were to live the Turkana lifestyle."

Religion also provided strength and a sense of purpose to individuals and groups. People said that churches and mosques were full after the natural disaster as people looked to God for wisdom and vigor.

Others said that the natural disaster renewed their commitment to help others, such as Erni, the midwife, in Aceh:

> God saved me because I'm still needed here. That's why my faith grows stronger and I become more loyal to Allah. If Allah has not wanted to take us away, there are reasons, there's a way to be saved. Just like me, I was under the water, messed up, but Allah said, "I don't want to take your soul away right now." That's what made me come to the barracks. I conclude that I have to do my job.

Solidarity with others was also a key factor in pulling through the crisis as a community. Being able to share and care for each other was a strength, and there are many stories of gifts of food, water, clothing, shelter, and helping each other rebuild. Activities such as praying or singing together also helped people to keep calm and hopeful. People did not want to be alone but grouped together to take shelter and rest. This solidarity does not divide people into victims and rescuers, but rather brings them together as equals working together for the common good. There were many stories of the disaster bringing people together in a shared humanity: people inviting people into their own homes and giving clothing to all comers, shopkeepers opening their doors and offering people goods, poor people providing tea and hospitality for incoming aid workers.

So, in every place, communities display resilience through agency, communication, cooperation, governance, identity, and solidarity. We can see that each of these factors are built through personal relationships. This brings us back to the idea of human relationships being systems embedded in systems. In a crisis, we see the value of each system—individuals initially reach out to family to make sure they are all right, and then to their social network or community/ies to find out what is going on and seek help or seek to help others.

Boy scouts in Pakistan, bus drivers in Aceh, custom (Kastom Gaden) gardeners in Solomon Islands, soap-opera actresses in Mexico, environmental activists in Burma, witch doctors in Kenya, and boat owners in New Orleans—the role these groups have in responding to natural disasters may not seem obvious, but each of these social networks made dramatic changes to the lives of people in crisis, as detailed in this book. Particularly in the first 48 hours of a disaster, these undervalued groups are more likely to save lives than official agencies tasked with rescues and relief. Resilience then is built on local relationships and the strength and value of these relationships to people affected by natural disasters for information, support, and action.

One surprise of our NVivo analysis is that people who have experienced natural disasters did not talk about a state that disaster and aid experts constantly talk about reaching: "Recovery." The word "recovery" occurs in the book chapters, but careful inspection will show that it is the interviewers who use the concept of recovery, rather than the interviewees, the people who speak from experience.

Listening to the voice of experience, we learn that disaster changes the community. There is loss of life and property, sometimes loss of a way of life. There is also a bonding through shared adversity, repeated storytelling, and coming together to commemorate the disaster. Conflict can occur at each stage, but particularly in the later stages conflicts arise within the community where resources have been distributed unequally or some people have exploited the situation. Also the new people and the agencies who enter the situation to help, initially welcomed with open arms, come to be seen more cynically over time. So, the process is one of adaptation to a new situation: what was has passed and what is normal is changed. For example, the contact with outsiders in Aceh, the transition from herding to farming in Kenya, the altered political balance in Pakistan, the relocation of villages in the Solomons. People can choose not to go back to what they had. In Pakistan we see that the people did not want to paint their houses the same color as they were before they fell down; their lives would not be the same after the earthquake and neither would their homes. In each case, the situation has changed and so have perceptions of what is "normal."

"Recovery" suggests a stage reached where it is like the patient is well and the illness never happened. But we see now that talking about "recovery" or "bouncing back" is, in community eyes, not the point. Instead, they talk about memory—how they remember and make sense of the disaster.

The study of memory as an individual cognitive process, necessary for learning, is an important area of psychology, but psychological studies of natural disasters have tended to focus on the negative role played by traumatic memories and the need to forget. The experience of sudden catastrophe was very vivid and remembered in detail, and in the case of these memories becoming highly intrusive, this might be seen as Post-Traumatic Stress Disorder (PTSD). Our method involved interviewing local community leaders and aid workers, helping them to recollect and tell their story and then reflect on community and wider responses to the disaster. While negative memories of a natural disaster might take the form of a PTSD and impair functioning, memories also have a positive role to play in terms of learning and adaptation. It was also apparent from our interviews that storytelling and sharing memories can strengthen connections and be a healing process for communities responding to a natural disaster. Living through a period of extreme danger and ambiguity can reinforce social bonds. In our healthy and capable respondents, the memories of extreme danger were vividly recalled "like it was yesterday" even when the people were interviewed four to six years after the event. The memories were painful but come to the surface of consciousness when intentionally recalled, rather than intruding themselves of their own volition, as happens in PTSD.

For people in Aceh and Pakistan, psychosocial or group activities that talked around or about the disaster were mentioned as important. In many cases, religion was a strong link to help people come together and cope; for example, in Solomon Islands churches gave sermons and led prayers about the tsunami. At the same time, some people sought religion for justification of disasters, to say that the disaster was punishment for immorality, and this kind of thinking could block adaptation in other ways such as through changing the community's relationship with the environment.

In other instances, rituals that were part of the local identity—a feast for Muhammed in Aceh or Mardi Gras in New Orleans—had a symbolic value in helping people feel together and bonded through local identities despite the disaster. In all cases, these ceremonies that connected individuals with their culture and community were mentioned more often than any official or new ceremony about the disaster as important. In the case of Turkana, they used their traditional way of storytelling of songs and special names for historic events to commemorate droughts. When people specifically remembered the disaster, they tended to do it in more personal ways to express themselves, such as the people in Burma who tattooed the date of Cyclone Nargis on their bodies or the graffiti people left on the damaged buildings in Gizo, Solomon Islands. People wanted to remember the natural disaster, not separately but together, in stories that integrated memory, learning, identity, culture, and personal and collective experience in a very powerful way of coping.

This sense of resilience being community capacities linked to a sense of community identity was described by peace researchers Lederach and Lederach (2010: 68–69): "Resiliency describes the quality needed to survive extreme conditions yet retain the capacity to find a way back to expressing the defining quality of *being* and the essence of a *purpose*." This quality of being and a sense of purpose is often rooted in local identity and culture, the shared ways of living and doing things together, which in a crisis provide some shortcuts in communication and common understandings about collective behavior.

So, since resilience is *everywhere*, what is the impact of external agencies—governments, armed forces, international agencies, the media—on resilience?

If resilience is built around personal relationships then we can see already that outsiders need to find a way to come into or to support these systems in order for their interventions to be fully appreciated by the local people. However, aid, while having the intention of helping the local people, in practice can have a "command and control" approach not dissimilar from governments and armies. This approach tends to focus on the establishment of camps and centralized distribution of supplies and services. In each place, people were eager to tell us about the errors of this approach—in Pakistan, we heard that the people did not want to go to the government-run camps, and that they wanted to stay on their land and resume their livelihoods, that the local NGOs felt that UN systems were not effective because they were imposed from above. In Aceh, the armed forces were seen as insensitive and the community leaders in the camps could not ask questions but just had to get back in line to receive supplies. In Solomon Islands one chief wanted a tent for the community but was given one that would only fit one family, and one village saw goods just being thrown off the boat without any questions about what the village needed and why. In Kenya, we heard that many NGOs did not consult the communities, that even food aid was seen as not helping local problems over the long term. In Myanmar, the set ways of the military and the international NGOs to do things contributed to the underacknowledgment of contributions from the local people and organizations in responding to the disaster.

The Western media too legitimizes this "command and control" approach, with its news adage of selecting "authoritative sources," generally state officials, instead of everyday people in a story. It is frequently only in desperation that reporters will interview the "man on the street," and instead they spend extraordinary amounts of time tracking down official sources and numbers.

The "command and control" approach, when it came to delivery of relief supplies and services, tended to be all about numbers—people are treated as individuals, and certain targets are set to give relief to a certain number of individuals. However, this individualistic view often clashed with local ways, which were more about getting a certain amount for the community and then sharing. For example, in Solomon Islands ownership of most things is communal, so village communities that received aid for the first time and were told some things were for only certain individuals found the idea strange, and it created conflict. Instead, aid agencies had to spend more to give to the entire community, to reach perhaps what they perceived as a needy few. The community's response to the natural disaster was collective and comprehensive, so there was some confusion when the aid response became orientated around individuals or narrow topic areas. Rashida Dohad described how her foundation's approach was to go into the villages in northern Pakistan, assess what was needed, and bring in housing, food, whatever was needed, but the UN "were doing it the other way around. They were saying "how much food?—"one cluster," "housing?"—"another cluster," and often they had no idea who was doing what." Then on a smaller scale, we heard from Chief Mana in Solomon Islands that big tents being supplied were only for schools leading him to ask: "Look, does the tsunami come and only affect schools or does it affect everyone?"

When people have experienced an event where nature overwhelmed them and they lost those things that people strive for over many years—houses, livelihoods, children—to not have control when external agencies come in the name of disaster relief can be upsetting, and in some cases lead to periods of feeling helpless and dependent. Aid workers and journalists who see inert survivors might assume helplessness but might actually be seeing people who have had a week of unimaginable horror and have finally felt sufficiently secure to take a pause in their frenetic activity. We might liken this to athletes who collapse at the end of a marathon. While it is true that runners do sometimes collapse at the end of a marathon, it would be misleading to only picture marathon races as consisting of athletes lying on the ground and struggling for breath. Most people rested when they needed to but were active; however, aid could encourage an abnormal response, as described by a religious leader in Solomon Islands: "I went around and saw people with broken tents six months after the tsunami and I said: 'What are you doing? Why don't you get up and work to help yourselves, do something now?' Most people it took them two days of rest because of the shock and then they started to rebuild. It was better to finish a house with local material than to sit around waiting."

For the people who coped, cooperated, and responded to the disaster there was also dismay that their efforts were not acknowledged by international agencies or the media. For many communities visited by international media, the media and the aid agencies were often seen as separate arms of the same foreign beast that had arrived in their midst offering help.

However, the outsiders came and began writing on a page that remained blank until they arrived rather than seeing communities which had already developed a story of resilience. Since the people we interviewed were active members of their communities, they unanimously shared this criticism of aid agencies in general—the lack of local control of what was happening or linkages to local activities and organizations already under way. Rashida commented about the dynamics in Pakistan:

> I think you should take the initial phase. I think the community was very together because they went through the whole tragedy together.... So, I think that initial phase brought people together. I think the distribution probably tore them apart, yeah. I think that the way the distribution was done, we could see the competition for goods which were insufficient to get to everyone... so in the hysterics of the moment, you probably you know grabbed for whatever you could get, but when you came back and you thought and you reanalysed your position and people around you, you know your family and friends and all that, so then you shared once again. And so because the social fabric was so well-knit, I think interventions tried to damage it but couldn't.... So... er that whole idea... that notion that when there is a disaster so the best thing for them for everybody is to come into camps and get fed and everything, I thought was completely missing the reality of people's lives, yeah?

On the other hand, the international agencies that funded organizations who already had links with local communities were empowering. Samina from Sungi in Pakistan described an example where a Norwegian funder allowed their funds allocated for a nonemergency purpose to be used to respond to the earthquake, enabling the organization to act quickly and locally. In Myanmar, the story of the establishment of the Local Resource Center to coordinate funds and efforts from international and local NGOs helped provide an effective response to the cyclone.

Having relationships with local people was more likely to help the agencies see the possibilities for the locals to do things themselves. Naila in Pakistan describes one case working with village women: "I said, *Amma* [literally translated as mother, but a respectful term used to address an elderly woman], 'this is your job now, it's your road, your village, now you do something about it. What is your committee, the village based organisation that we had developed together doing about it'? She said: '*Baitay* [literally translated as child, but an affectionate term used to address a much younger person], we'll do something, we'll find some way.' Believe me, then the community people contacted their local council member and the mayor and the vice mayor and they cleaned and opened up the road. This had never happened before."

Similarly, agencies that helped people return to their livelihoods or create new sources of income were seen as valuable and practical for communities—such as farming and irrigation in Kenya, support to start small women's businesses in Aceh, and many others.

While there was gratitude for assistance there was also comment on the chaos and lack of coordination of aid, the inappropriateness of some donations, and the waste caused by mismatches. The process of how aid was given was in many ways more important than what was given when people were thinking about its impact on community resilience.

Once there is the onset of a disaster, there is no time for outsiders to learn about the local culture, build links with local organizations, and deliver aid that meets people's needs. There is only time for crisis management, and it often takes time to mobilize relief efforts and overcome logistic problems. So, it is extremely difficult for new organizations to come into a local area and respond to the crisis promptly. Journalists coming in from outside find similar pressures and often end up misreporting the situation because they do not understand what they are seeing, as a local New Orleans *Times-Picayune* journalist John Pope described: "A crisis is not time to be making new friends. You need to establish your sources as you go along and get their cell phone numbers and that is how you get to them....For two months (after Hurricane Katrina) I felt like the man who sweeps up after the elephants at the circus [the international and national reporters coming in from outside]. I was saying in essence: 'Just calm down'" (Deutsche Welle Global Media Conference, 2010).

Similarly, an international aid agency's "need for speed," often driven by their large fund-raising campaigns, creates undue pressure to establish a relationship between international agencies and the local community and sometimes replicates efforts to solve problems already made by local communities. The "need for speed" and the idea of "recovery" lend themselves to seeing each disaster as a separate crisis that is dealt with, and then the attention of the international community moves on to another crisis elsewhere in the world. In actuality, certain areas are prone to certain types of disasters and more could be done to build ongoing links with local organizations and work on prevention of and preparation for disasters.

The crisis management model also neglects the later stages of disasters. A number of local people stressed that aid came in the form of handouts, say of food, which had a limited value. Putting people into tents and barracks might allow for efficient handing out of food, but was resisted by local people, whose longer-term recovery prospects involved staying with their land and growing their own food. That is, the local people could see the importance of adaptation, but this potential was rarely fulfilled.

International experts tend to be expert in disasters, rather than experts in language or local culture or community development. While there are commonalities between an earthquake in one location and an earthquake in another location, managing the disaster needs to take much more account of the cultural, economic, political, and social dimensions that may be associated with each location.

Our point here is that aid will have lasting as well as immediate effects and that agencies should look not only at timely intervention in a crisis, but also at longer-term relationship building. The history of colonization is replete with examples of the destabilizing effect on local relationships when colonial powers build up the power of one group vis-à-vis another or otherwise changed permanently community life through its influence.

Natural disaster responses from external agencies bring together people who do not normally come into contact with each other—local communities and armies, local communities and foreigners. In fact, all of the major disasters described in this book were examples of areas not normally visited by foreigners or of aid on the scale that was provided after the disaster. Similarly, natural disasters are one of the few chances the rest of the world gets to see some places through the eyes of the media—the Pew Research Center for People and the Press (as detailed in King, 2002) states that stories about natural disaster rank second (after terrorism) as being the most popular media stories. Given television's news prime focus when selecting stories on choosing dramatic images, natural disasters are featured every day. With such prominence being given to natural disasters in the media, the coverage can shape outsiders' perception of the entire nation affected by the disaster. Consumers of media will forever associate the name

Aceh with tsunami, Ethiopia with famine, Haiti with hurricane. The places are defined in popular consciousness by media coverage of what people do in a severe crisis.

These times of interaction between the local communities and the international community were opportunities to build better relationships, to "open-up" people that have been living insular lives to new people and ideas. In some cases this happened through better education in remote areas and changes in the role of women, with NGOs encouraging women in leadership roles. In other cases, interaction with outsiders caused the local people to distrust and "close ranks" from outsiders because aid was unequally distributed, given in ways that undermined dignity or showed a lack of appreciation for the local culture. From examples as diverse as supplying Father Christmas dolls in the Islamic province of Aceh to not consulting indigenous chiefs in Solomon Islands, understanding the local culture and the local language was seen as critical if the external agencies were to have good relationships with the local communities. Even the sort of data gathered in the chapters of this book would not have been possible without the trust already established between the local researchers (who in most cases were not community members, not international workers but somewhere in between). The researchers understood very well the norms of the community and worked within these to solicit feedback from the people interviewed. The narrative and discussion style of interviews allowed for data that would not normally be collected in, for example, an aid-project monitoring and evaluation report, a news bulletin, or a topic-specific research survey. It allowed the people to talk in general terms about the impact of aid on their economy, politics, natural environment, local community, culture, and outlook.

Local people talked about a set of relationships nested outside that of aid-giver and aid-receiver, which are dramatically affected by the arrival of aid. These are often very visible to the locals but unintended or ignored by the external agencies.

One is the economic consequences, through the arrival of foreign staff, goods, and organizations. When local people learn that outside aid workers and experts are highly paid and live an affluent lifestyle, they are less likely to see them as heroic rescuers and more likely to question the economic inequalities and underlying motives. Aid changes the local economies—in some cases prices rise, in other cases staff in local NGOs are pulled into international organizations by high salaries, in many cases the price of rent goes beyond what most local people can afford. These changes, even small ones, such as the arrival of a chainsaw in a Solomon Islands village, create new "haves" and "have-nots," giving birth to new elites or strengthening old ones.

Then there are the political ramifications. National and local governments also have a complex relationship with the local and international community during the aftermath of natural disasters. On the one hand, the bypass of the government was seen as a problem—in Solomon Islands the direct donation of money to politicians and not to the agency charged with dealing with disasters led to conflict and corruption. On the other hand, sometimes the focus of aid agencies on working through the government was seen as disempowering communities, and middle people between the community and the aid could misuse funds and create lack of transparency.

Aid agencies could be seen as giving mixed messages in Myanmar: on the one hand they needed government permission to come inside the country, but then in practice they tended to act autonomously in the conduct of their programs. Whereas the local people built relationships with the army so that they could continue to work with local communities, the international agencies often did not see the value of dialogue. The role of the media was seen as counterproductive by locals working to change the very issues (lack of freedom, development) that the media highlighted through representing people as without agency due to oppression. One interviewee described the UN as perpetuating this impression of the community as inactive when in fact they were just "silent."

A great deal of media attention was directed at the Myanmar government's position in relation to accepting, or refusing to accept, international humanitarian assistance. This media focus forms part of a trend that sees news stories about Myanmar narrowly focus on the brutality of the military regime and deep divisions between Myanmar people of different religious and ethnic backgrounds, emphasizing its isolation within the international community. It is important, however, that along with the stories of horror and destruction, there is room given to acknowledge and explore the positive and negative parts of this tragedy. In critically engaging with the impact of Cyclone Nargis, we need to go beyond simplistic pronouncements. We must unpack inherent complexities to understand the Myanmar context and to trace its changes and shifts.

Where there are good relationships between local communities and state agencies, the solidarity and increased communalism with which local communities respond to a disaster can spread out to an increased sense of nationalism. However, closer contact might further erode negative relationships. In a disaster, the state will be personified to local communities through the conduct of visits by state officials, the presence of emergency services, and the armed forces. Later, as the disaster scenario unfolds, a more bureaucratic aspect will become salient, with the distribution of aid and the regulation of reconstruction impacting on local communities, sometimes in contrast to the earlier high speed of event and response, in an infuriatingly slow manner.

Once aid agencies make their entry, there seems to be a marked lack of cooperation among them. Instead of working together to help repair the situation, the agencies are seen to be in rivalry and competition. Individuals, groups, and organizations that could respond to local needs promptly and appropriately gained in standing—for example local church or community organizations that were able to respond in this way gained in stature. If these groups in turn support particular political factions such as the Taliban, then the disaster serves to enhance their power in the country.

These economic and political consequences of aid were left behind after the aid programs were finished and left. The time-limited involvement of many NGOs meant that people were often wary of new ones; as local woman Ereng said in Kenya, NGO people were seen as "temporary people." The NGOs did not have the "shadow of the future" in some local communities (which researcher Robert Axelrod described as the key to cooperative relationships). But the local community would live with each other and the situation permanently.

What about the Worst of Times?

Tahira Abdullah, rights activist, from Pakistan saw a truism in her experience: "What was it that Dickens said in...in... *Tale of Two Cities*? 'It was the best of times and it was the worst of times.' We've seen the worst aspect of human nature."

It might seem that our analysis has presented a rather rose-colored view of community life. Where do the media stories of people exploiting the situation come into this analysis: the looters, the profiteers, the rapes in refugee camps, the outbreaks of group violence that take advantage of the chaos to exploit the situation?

Part of this perception is due to selective reporting with a focus on drama and a lack of understanding of the local communities. Imagine, you are a television journalist who arrives at a disaster scene in another country, there is a mother sitting on the ground holding a dead baby and wailing in distress, there is a group of young men talking animatedly to an older man, there is a group of men running with a dead body, there is a young girl walking down the road with eight small children trailing behind her with various injuries, there is a group of policemen driving past slowly in a car. What are you going to film, and who are you going to interview? How do you make sense of the community at a time like this? Chances are, if you are trained as a journalist, you will film the woman wailing, and the group of men running with the dead body. They are the most dramatic images that will draw the viewer to the story and fulfill the old news editor's motto, "If it bleeds, it leads [the news]." And you will interview a policeman as he is a representative of the state and in journalistic tradition therefore seen as an "authoritative source" on what is happening.

So, cut the story together with the wailing woman, dead body, and policeman, and what does it tell the viewers about what people are doing in the disaster? It tells them that the people are distressed, people are dead, and the police are there to act. It is a story with all questions answered.

But what about all the other stories at the scene and all their unanswered questions? The young girl with the children could be taking them to a clinic (as in Pakistan), the older man arguing with the youth could be organizing them to go out and find food in the rubble (as in Solomon Islands), the woman wailing could be a midwife who moments later got up and realized she had to go help other women in need of health services (as in Aceh), the group of men could be running to take the body not to a morgue or hospital but to a volunteer-organized Free Funeral Service (as in Myanmar). If any or all of these stories were told, how would this change the viewer's perception of this group of people in a crisis?

A story where people compete individualistically for more power and goods, through violence or theft, makes dramatic visual images and good television drama. However, most of us will not play this part—when faced with a natural disaster, the majority of the people act according to the pattern of relationships more likely to lead to their personal and collective survival and growth—they stick together, communicate about their problems and possible solutions, and help each other out. The majority of people in a crisis are not disruptive but resilient.

So how do we tap this resilience, this kind of cooperation, agency, communication, governance, identity, and solidarity that binds people together in bad times?

Many people will say that natural disasters "brings out the best and the worst in people." Given the trying circumstances, community members will hope for the best and justifiably feel shocked and betrayed if the worst comes out. But natural disasters are a strange and unusual times—the individual pressures of everyday life are off and instead there is a common threat to all—so altruism and community spirit is high. It is far more difficult to mobilize people around ongoing problems that lead to the worst in people, such as poverty, oppression and exclusion of women and minorities, mental illness, and environments of crime. As time goes on, cooperation will fade and these ongoing problems will come back, often revealed in a new light. In truth, New Orleans was statistically a city with areas of high crime and poverty and low education before Hurricane Katrina came along—so the disaster did not bring up new problems, instead it revealed what was *already there*. Then the challenge was to try and use the disaster as an opportunity to press for change.

Theories of peace and conflict distinguish between conflict, which can be constructively resolved, and violence, which is morally unacceptable. Violence can take the form of direct physical abuse, but it can also take the form of structural inequalities or cultural discrimination and exclusion (as detailed by Galtung, 1978). Structural violence is built into the fabric of society and kills not by commission but by omission and exclusion. Cultural violence silences some voices of some people and privileges others. In a disaster, these three forms of violence can be seen to play out.

The potential for cooperation and constructive responses to a natural disaster are undermined by inequitable power relationships. The impact of a natural disaster is not evenly spread, with most disasters counting more women and children among their victims. While it is of course true that the physical size and strength of healthy adult men is an advantage when facing physical adversity, especially compared to babies who might be washed from their mothers' arms during a tsunami or killed by falling debris that would merely injure an adult, this is not the whole story.

Take for example, Pakistan, where disadvantages of women were not merely a matter of physical strength. Because of the social arrangements, women were more likely to be confined indoors and hence more likely to be trapped in falling buildings. Also women are the caregivers, so some who had escaped safely then went back into buildings to fetch elderly or ill relatives, exposing themselves to danger again. Pakistani women were particularly at risk and exposed to abuse in camp settings, and there were also reports of male leaders barring sanitary supplies sent by aid agencies because they thought such supplies would make women immoral. Women were not free to participate in decisions, and community members were dominated by men. There are also exceptions to this picture, with girls in Pakistan showing great ingenuity and resilience; however, the story of women having a greater burden than men is one found in many of the places visited in this book.

Then there are issues of economic and political power on an international scale—for example, what language the warning is in. Who does it reach? For the Asian tsunami the warning reached Western capital cities, but not the Acehnese. Even after the extreme devastation of the tsunami in 2004 and increased investment in early warning systems, the tsunami warning in October 2010 did not reach Indonesia, because the buoy closest to Indonesia was not functioning. So, we see that imbalances in who has information and how knowledge is transmitted have not changed, despite the disaster.

Similarly, in our interviews we can see that it is those working in NGOs who received early warnings. Within a country it tended to be the more affluent city dwellers who had television sets to receive news, mobile phones to call, and consult others, while rural workers in remote areas were less likely to hear reports of impending disaster. These people would rely more on folk knowledge to predict weather patterns and crises, and whatever previous experience they had of disasters.

While we call them "natural disasters," to some extent none are entirely natural as we live in a world where humans interact with nature and themselves, and these relationships generally tend to be at the expense of the poorest humans. People in the developing world are four times as likely as people in the developed world to die in natural disasters particularly floods, according to a report by Action Aid. Urbanization aggravates flooding by restricting where waters can go, and structures in slum areas are often particularly vulnerable to shocks (*The Futurist*, July–August 2007: 6). The interplay of human and environmental factors is well illustrated by the Turkana drought, where patterns of agriculture and animal grazing are tied up with local identity and tribal relationships, such that the effects of the drought have been exacerbated by overstocking of cattle. Indeed, some biologists, such as Tim Low (2002), suggest that part of the problem is in seeing humans as being somehow *outside of* or *in opposition to* nature. Instead, there is increasing stress on the need for a more intelligent and harmonious approach to relationships between people and with the environment.

When responding to natural disasters with emergency relief, some people are more likely to get both more relief and more quickly. Consider, for example, firefighters in California who have access to airports, roads, vehicles, and supplies to conduct large operations and will do so using common procedures and a shared language. In the remote NWFP of Pakistan the terrain is rugged and difficult. The location of villages in the mountains is a matter of local knowledge rather than Google maps. Even the cultural distance between these rural people and helpers from Islamabad who speak their language is something to bridge, while the gap between the locals and incoming foreigners is vast.

Natural disasters highlight instances of structural violence and exclusion inbuilt in the system of the way things are and the way they are done, which can impact on how a community responds to a disaster and how agencies who come to help respond to that community. Bringing this into the spotlight can kick-start both authorities and locals into addressing problems.

The media can also be used constructively by communities to highlight these inequities and apply pressure on the authorities for change, as radio journalist Adrian Ginia described:

> All the NGOs started using our program for awareness programs about health and all these things and we had news coming in from people about supplies and arguments. The main issues [people had] were distribution, some things didn't make it to their destination, they got lost along the line. Even the NGO workers, they ended up with the most benefit from distribution without assistance reaching those who were really victims. All these things were said [on the radio] they came in different forms but there were complaints about distribution of supplies. Others said 'why not us?' 'Why is World Vision doing something in this community and not our community?' On the other side, the Red Cross, Oxfam etc have their own stages of the process— emergency relief, rehabilitation and reconstruction and they tried to make people understand.

Natural disasters can raise expectations of communities and prompt them to take action against what they see as "the worst in people." Local people who exploit aid initiatives are likely to incur bitter resentment from their peers in the community, all the more so because of the social bonding and altruism that has occurred, and the openhanded generosity and the brave sacrifices made by many people in the period of acute danger. Similarly, it makes very little sense for any supermarket chain to hoard and guard their produce in a time of disaster: openhanded distribution makes very good sense not only morally, but also in terms of good business advertising and public relations, as realized by giant retailer Wal-Mart, which opened its doors in New Orleans.

Good politicians too seem to intuitively understand the dynamics of power and are quick to make their way to the site of a natural disaster, to express solidarity with the local people. While this may seem like window dressing to the cynical outsider, discussions with survivors suggest that it is very important and leaves them feeling well supported. Even if there is logically nothing accomplished by their physical presence, it psychologically and emotionally extends the sense of solidarity and reinforces the idea of the community being supported by the state at a crucial time. Similarly, proper representation of the state at funerals, commemorations, and memorials can reinforce solidarity between the community and the state. Where this is neglected, perhaps because the affected area is poor, remote, or not politically persuasive, unique opportunities for building harmonious relationships have been missed.

Similarly, the Western media will in general interview its own national citizens and aid workers in a natural disaster in the developing world. These eyes become the general viewing public's eyes on the natural disaster and on the communities affected. Building images of Westerners as rescuers serves to disempower and misrepresent communities and also sometimes to reinforce the idea that police and armies are necessary, that peace is maintained through armed force. One of the basic assumptions implicit in media reports is that it is the aid agencies and the governments that are the primary response to the disaster, whereas we know it is the communities themselves that have been responding and need help. A greater focus on what community members do could help them feel acknowledged and valued, and also influence the mind-set of those in power in the relief camps, the government agencies, and those donating money from afar to support community resilience rather than dependency.

Training for psychiatrists, psychotherapists, clinical psychologists, and counselors has to address the allure of seeing oneself as a rescuer and the dangers of creating dependence. Perhaps there is room for some transfer of knowledge into the training of aid workers and disaster experts. Not that aid workers and disaster experts should come to see themselves as therapists, but perhaps there are some techniques that would encourage their being reflective about their own role as helpers and more aware of the limits to their own understanding.

Delivery of aid that is timely, that meets the needs of the local people, and that is seen as done with them and not for them will create a reservoir of social capital will build loyalty and good will. It is a unique opportunity to create new bridges of understanding and can establish a relationship that is deeply enduring. We heard in Pakistan from Naila that women "could not have even imagined" how they would benefit from aid through education and being involved in what before would have been considered men's business, such as building a road.

Interventions can help support cooperation and inclusion rather than competition in communities. People who ordinarily may not work together start to cooperate to deal with common problems, or one group can help out the other group as it sympathizes with their loss, creating bridging social capital between different groups. For example, in Myanmar people talked about how the different ethnic and religious groups worked together often for the first time on disaster relief and development, how people "didn't see religion and race, they just saw human beings."

Aid programs can help communities take back control of their lives, to govern themselves, and find collective solutions to problems: committees are called together and new ones developed, traditional leadership plays a role, education offers new solutions, and different groups of people come together faced with the common challenge. The aid projects remembered as worthy and effective tend to build in development to relief as Chris, pastoralist and NGO worker, in Kenya said: "Whatever you do in relief time, you should actually build into development work. And that link should really be fortified."

A changed understanding and ability to adapt to be better prepared for disasters can be a result of aid workers' communications and practical assistance: in Kenya and Solomon Islands we saw that people took up the encouragement and training to grow food, for example.

External agencies can support the identity and solidarity of communities by working through local culture rather than working over the top of it: such as through group psychosocial activities to talk about the disaster in Aceh and Pakistan.

And all this can be recognized better in the reporting and analysis of the media and the general knowledge of the donor. The media can portray a community as active rather than passive, as having agency rather than being helpless, that they are survivors rather than victims, coping rather than traumatized, strong rather than weak. This is what the media tends to do for its own hometown—after Hurricane Katrina, Louisiana-based local newspapers were three times more likely than other US papers to present page-one stories about the disaster in a positive light. Affected by the natural disaster themselves, journalists, photographers, and editors had a greater focus on survivors, were less likely to blame them, and be negative about the place and people. Instead they gave locals a voice—interviews with the common "man in the street" increased from 38 percent to 63 percent from the year before Hurricane Katrina to the year after in local newspapers (Roberts, 2010: 61). As Roxanne K. Dill from the Manship School of Mass Communications at Louisiana State University mused, local newspapers framing things in this way served an important role for the community: "Perhaps their local newspapers sought to foster hope by reminding readers of their own innate resiliency while preserving one of the community's important information lifelines" (Roxanne K. Dill in Izard and Perkins, 2010: 43).

While the natural disaster brings people together, aid need not tear people apart. What is required is a more holistic approach to the formation and delivery of aid and the feedback of its effects into aid agencies for learning. In evaluating interventions such as responses to natural disasters it is important to look beyond simply achieving the project goals determined by the agency. Vivienne Jabri (1996), in her study of discourses on violence, says that actions have unpredictable as well as predictable effects. The impact of aid on the local human ecology needs to be considered, and a deep consideration will have to involve listening to local voices. Greater use of conflict analysis and resolution techniques by aid workers that stress the importance of listening and understanding could be a helpful influence on the way things are done.

Despite all the criticisms of aid the local people had, few talked about doing away with it altogether. Instead they talked about a dysfunctional relationship they wanted to change to make it more like the family and friendship relationships people rely on in a natural disaster—built on understanding and familiarity, sharing knowledge in conversation, practicality and the needs of daily life, joint goals for the future, and a sense that their identity and dignity were enhanced through the interaction. It was this kind of relationship, whether with themselves or with other groups and organizations, that people felt made their communities more resilient.

Our journey to gain a better understanding of community responses to natural disasters has been an outward one, taking us to different regions of the world, and with the assistance of our local researchers to converse with some extraordinary survivors.

We took our experience as peace and conflict researchers to examine how external interventions can promote peace within a community and between itself and others. So, we found that cooperation and solidarity are high in the response to a natural disaster, that a common threat seems to create peaceful relations between people. The local NGOs that arrive are also part of this peaceful movement to grow from the crisis and build enduring relationships. However, as the media puts the spotlight on the local community and more distant organizations arrive to work there, with less understanding of the local communities, sometimes less respect, ways of doing things that create conflict, competition, and hostility or resentment are established. The "command and control" approach promotes victims, not resilient survivors—people who want to get on with their lives clash with organizations, promoting a "handout mentality" that inevitably comes with large-scale aid. To add to the tensions, the media will report the negative elements of this (the snatching of supplies, the opportunists, the dependency). Then the media and the international organizations leave, but the lives of the local people and their challenges remain. Local organizations will be left to develop sustainable and cooperative relationships between themselves and the communities and within communities themselves. The community resilience seen so clearly in the immediate response to the disaster may take a long time to emerge out of the schisms, confusion, and mistrust left behind when the outsiders move on to the next crisis.

While our study is global, it also brings our thoughts back home and challenges us to reflect more on our habitual dichotomized ways of thinking and tendency to form stereotypes. Our travels have demonstrated that poor people in remote regions in developing countries are not passive victims of their fate, waiting for us (or the agencies we give donations to) to come and rescue them, but are capable and creative communities.

Of course, resilience is not a matter of being invulnerable, and one danger in using the term is that it may be interpreted to justify neglect or abuse. If individuals or communities are seen as being able to bounce back from any blow, or like steel even strengthened and tempered by fire, then there is a danger that they will be treated harshly. All human beings are fragile, and some disasters are going to overwhelm even the most resilient and well-resourced communities.

Community resilience, the quality of being able to enhance chances of communal survival through effective collaborative decision making and local action, is everywhere, but it is not a guarantee of survival. The concept of community resilience should instead be used to inculcate respect for the strength and integrity of local communities. Our new understanding of what communities can do when facing natural disasters should not make us turn away, unconcerned about the dangers and problems they face. Instead, the stories of survivors make us realize that one day we may all face a natural disaster and look around to see who is by our side—who can help us find ways to help ourselves.

References

Atton, C. (2007), "Keeping the Peace: Media Representation of the Anti-Gulf War Movement in the British Press" in S. Maltby, and R. Keeble (eds.), *Communicating War: Memory, Media and Military,* Arima: Suffolk, UK.

Brinkley, Douglas (2006), *The Great Deluge: Hurricane Katrina, New Orleans and the Mississippi Gulf Coast,* New York: Harper Collins.

Carson, R. (1962), *Silent Spring.* Boston: Houghton Mifflin.

Christie, D., Wagner, R., and Winter, D. (eds.) (2001), *Peace, Conflict, and Violence: Peace Psychology for the 21st Century,* Englewood Cliffs, NJ: Prentice Hall.

Deutsche Welle Global Media Conference, Bonn, June 21–23, Dart Centre moderated panel transcript as found on: http://dartcenter.org/content/transcript-witnessing-human-cost-climate-change

Dill, R. K. (2010), "Local Coverage: Anticipating the Needs of Readers" in R. Izard and J. Perkins (eds.), *Covering Disaster: Lessons from Media Coverage of Katrina and Rita,* New Brunswick USA: Transaction Publishers: 39–54.

Drabek, Thomas E. (2010), *The Human Side of Disaster,* USA: CRC Press, Taylor and Francis Group.

Galtung, J. (1978), "Peace and Social Structure," *Essays in Peace Research Volume 3,* Copenhagen: Ejlers.

Jabri, Vivienne (1996), *Discourses on Violence: Conflict Analysis Reconsidered,* UK: Manchester University Press.

King, D. C. (2002), *Selling International News,* Massachusetts: Harvard University.

Lederach, J. P. and Lederach, A. J. (2010), *When Blood and Bones Cry Out: Journeys Through the Soundscape of Healing and Reconciliation,* Brisbane, Australia: University of Queensland Press.

Low, T. (2002), *The New Nature,* Australia: Viking.

Poniatowska, E. (1988), *Nothing, Nobody: The Voices of the Mexico City Earthquake,* Philadelphia: Temple University Press.

"Rising Waters, Drowning Hope," *The Futurist,* July–August 2007: 6.

Roberts, S. (2010), "Split Personalities: Journalists as Victims" in R. Izard and J. Perkins (eds.), *Covering Disaster: Lessons from Media Coverage of Katrina and Rita,* New Brunswick USA: Transaction Publishers, 55–70.

Suzuki, David (1997), *The Sacred Balance: Rediscovering our Place in Nature,* Vancouver: Allen and Unwin.

INDEX